PRINCIPLES OF POLYMER PROCESSING

Front Illustration
85 mm diameter single screw extruder for thermoplastics; fitted with a crosshead
die for wire covering; the barrel is electrically heated and air cooled.
Reproduced by kind permission of Francis Shaw & Co. Ltd, Manchester.

Principles of Polymer Processing

ROGER T. FENNER
Ph.D., B.Sc.(Eng), D.I.C., A.C.G.I., C.Eng., M.I.Mech.E.,
Lecturer in Mechanical Engineering,
Imperial College of Science and Technology,
London

First published 1979 by
THE MACMILLAN PRESS LTD
London and Basingstoke
Associated companies in Delhi Dublin
Hong Kong Johannesburg Lagos Melbourne
New York Singapore and Tokyo

Typeset in 10/11 Press Roman by
Styleset Limited, Salisbury, Wilts.
and printed in Great Britain
by J. W. Arrowsmith Ltd, Bristol

British Library Cataloguing in Publication Data

Fenner, Roger Theedham
    Principles of polymer processing.
    1. Polymers and polymerization
    I. Title
    668.4   TP1120

    ISBN 0−333−25527−5
    ISBN 0−333−25528−3 Pbk

Contents

Preface   vii

Notation   ix

1 Introduction   1

1.1 Polymeric Materials   1
1.2 Polymer Processing   2
1.3 Analysis of Polymer Processes   2
1.4 Scope of the Book   3

2 Introduction to the Main Polymer Processes   4

2.1 Screw Extrusion   4
2.2 Injection Moulding   9
2.3 Blow Moulding   13
2.4 Calendering   13
2.5 Other Processes   15
2.6 Effects of Processing   15

3 Processing Properties of Polymers   16

3.1 Melting and Thermal Properties of Polymers   16
3.2 Viscous Properties of Polymer Melts   17
3.3 Methods of Measuring Melt Viscosities   20
3.4 Elastic Properties of Polymer Melts   29
3.5 Temperature and Pressure Dependence of Melt Properties   31
3.6 Processing Properties of Solid Polymers   32

4 Fundamentals of Polymer Melt Flow   33

4.1 Tensor Notation   33
4.2 Continuum Mechanics Equations   35
4.3 Constitutive Equations   37
4.4 Boundary Conditions   43
4.5 Dimensional Analysis of Melt Flows   43
4.6 The Lubrication Approximation   45
4.7 Mixing in Melt Flows   49

5 Some Melt Flow Processes   53

5.1 Some Simple Extrusion Dies   53
5.2 Narrow Channel Flows in Dies and Crossheads   66
5.3 Applications to Die Design   74
5.4 Calendering   79
5.5 Melt Flow in an Intensely Sheared Thin Film   85

6 Screw Extrusion   93

6.1 Melt Flow in Screw Extruders   94
6.2 Solids Conveying in Extruders   115
6.3 Melting in Extruders   123
6.4 Power Consumption in Extruders   136
6.5 Mixing in Extruders   137
6.6 Surging in Extruders   138
6.7 Over-all Performance and Design of Extruders   139

7 Injection Moulding   145

7.1 Reciprocating Screw Plastication   145
7.2 Melt Flow in Injection Nozzles   147
7.3 Flow and Heat Transfer in Moulds   152

Appendix A   Finite Element Analysis of Narrow Channel Flow   160

Appendix B   Solution of the Screw Channel Developing Melt Flow Equations   163

Appendix C   Solution of the Melting Model Equations   168

Further Reading   169

Index   173

Preface

The increasing use of synthetic polymers in preference to metals and other engineering materials for a wide range of applications has been accompanied by the development and improvement of processes for converting them into useful products. Indeed, it is often the comparative ease and cheapness with which polymeric materials can be processed that make them attractive choices. Because of the relatively complex behaviour of the materials, polymer processes may appear to be difficult to understand and analyse quantitatively.

The purposes of this book are to introduce the reader briefly to the main methods of processing thermoplastic polymers, and to examine the principles of flow and heat transfer in some of the more industrially important of these processes. Much attention is devoted to the two most widely used methods − screw extrusion and injection moulding. Quantitative analyses based on mathematical models of the processes are developed in order to aid the understanding of them, and to improve both the performance and design of processing equipment. In addition to algebraic formulae, some worked examples are included to illustrate the use of the results obtained. In cases where analytical solutions are not possible, methods of numerical solution using digital computers are discussed in some detail, and typical results presented.

This book is partly based on courses given by the author to both undergraduate and postgraduate students of mechanical and chemical engineering at Imperial College. The level of continuum mechanics and mathematics employed is that normally taught in undergraduate engineering courses. Although tensor notation is introduced for conciseness in presenting general continuum mechanics equations (in chapter 4), no prior knowledge is assumed, and it is not used in the later practical applications. The book is therefore suitable for engineering undergraduates and postgraduates, and other students at equivalent levels. Practising polymer engineers may also find it useful.

The author wishes to acknowledge the contributions made by many colleagues, students and technical staff to the work on polymer processing at Imperial College, on which this book is firmly based. Much of the material concerned with cable-covering-crosshead design presented in chapter 5 is drawn from the current work of Mrs F. Nadiri. The very skilful typing services of Miss E. A. Quin are also gratefully acknowledged.

Imperial College of Science and Technology,
London

ROGER T. FENNER

Notation

The mathematical symbols used in the main text are defined in the following list. In some cases, particular symbols have more than one meaning in different parts of the book. Where appropriate, the chapter or section in which a particular definition applies is indicated in parentheses. Many of the symbols used in the appendixes are the same as in the main text, although some new notation is introduced and defined within each appendix.

A	dimensionless coordinate in direction of flow (section 4.5)
A	a function defined in equation 5.101 (section 5.5.1)
A	cross-sectional area of a screw channel (section 6.2.1)
A	integration constant in equations 5.6 and 7.5
a	radius of a cone-and-plate-rheometer cone
a_1, a_2, a_3	constants in general thermal boundary condition equation 4.47
B	integration constant in equation 5.35
B	a function defined in equation 5.101 (section 5.5.1)
Br	Brinkman number
b	radius of drum driving cone-and-plate rheometer (section 3.3.1)
b	temperature coefficient of viscosity at constant shear rate
b_i	body force vector
C	consistency (viscosity) in power-law equation
C	integration constant in equation 5.102 (section 5.5.1)
C	a function of screw geometry in equation 6.79 (section 6.2.1)
C_p	specific heat (of melt)
C_{pm}	specific heat of a melt
C_{ps}	specific heat of a solid polymer
$C_1 - C_5$	constants defined in equations 5.5, 5.20, 5.32, 5.40 and 5.81
C_1, C_2	constants defined in equations 6.92 and 6.96
c	radial clearance between barrel and flight tips
D	diameter of a circular flow channel, including a rheometer capillary
D	internal diameter of an extruder barrel (chapter 6)
D'	diameter of a molten extrudate after leaving a capillary
D_0	reference mould cavity diameter
D_1, D_2	initial and final diameters of a tapered circular die
E	a function defined in equation 5.37 (section 5.1.3)
E	a function defined in equation 5.106 (section 5.5.1)

E	a function of screw geometry in equation 6.79 (section 6.2.1)
$\dot{E}$	rate of working per unit area of extruder-barrel surface
E_z	power consumption per unit downstream length of channel
$E_z{'}$	power consumption per unit downstream length of flight
E_1	power consumption per unit length of upper calender roll
e	capillary end correction (section 3.3.3)
e	rate of extension of a filament (section 4.3.4)
e	width of a screw flight (chapter 6)
e_{ij}	rate-of-deformation tensor
F	force on a calender roll, per unit length of roll (section 5.4.1)
F	net transverse force at the flights in a screw-feed section (section 6.2.1)
F_D	drag flow rate shape factor
F_P	pressure flow rate shape factor
f_D	drag flow velocity shape factor
f_P	pressure flow velocity shape factor
G	Griffith number
Gz	Graetz number
G_1, G_2	functions introduced in equations 6.84 and 6.85
g	summation counter used in equations 6.46 and 6.47
g_{ij}	velocity gradient tensor
H	depth of a flow channel, including screw channel and mould cavity
H_s	depth of solid bed
H_0	distance between the lips of a flat film die (section 5.3.1)
H_0	minimum distance between the rolls of a calender (section 5.4)
H_0	initial thickness of a thin melt film (section 5.5.2)
H_0	initial screw-channel depth (section 6.1.4)
H_0	reference-mould-cavity depth (section 7.3.3)
H_1	initial channel depth in a flat film die (section 5.3.1)
H_1	distance between calender rolls at point of separation (section 5.4)
H_1, H_2	initial and final depths of a tapered flat-slit die (section 5.1.2)
H_1, H_2	functions introduced in equations 6.84 and 6.85
h	height of a flow element in a cone-and-plate rheometer (section 3.3.1)
h	heat-transfer coefficient (section 4.4)
h	length of a striation (section 4.7)
h	thickness of solid skin in a mould cavity (section 7.3.4)
I_1, I_2, I_3	principal invariants of the rate-of-deformation tensor
$I_2{}^*$	dimensionless form of the second invariant
K	ratio between outer and inner radii of a flow channel (section 5.1)
K	a function introduced in equation 6.78 (section 6.2.1)
k	thermal conductivity (of melt)
k_m	thermal conductivity of a melt
k_s	thermal conductivity of a solid polymer
k_1, k_2, k_3	ratios between compressive stresses at barrel, channel sides and screw root, respectively, and stress in downstream direction

L	length of a flow channel, including rheometer capillary
L	axial length of a screw
L'	length of flow channel required for fully developed flow
L_0	distance between pressure initiation and nip in a calender
L_1	length of the die lips at the centre of a film die (section 5.3.1)
L_1	distance of point of separation from calender nip (section 5.4)
L_1, L_2, L_3	lengths of three rheometer capillaries (section 3.3.3)
l	fluid displacement in mixing analysis (section 4.7)
l	screw-flight pitch (chapter 6)
M	degree of distributive mixing
M	a function introduced in equation 6.80 (section 6.2.1)
$\overline{M}$	mean degree of distributive mixing
$\overline{M}_x$	mean distributive mixing per unit channel length
M_{mz}	downstream mass flow rate in the melt pool
M_T	total downstream mass flow rate in a screw channel
m	number of screw channels in parallel (number of starts)
$\dot{m}$	total mass flow rate from an extruder
m_{fx}	leakage mass flow rate per unit downstream distance
m_1	resultant upper melt film mass flow rate per unit width of film
m_{1x}, m_{1z}	upper melt film mass flow rates in x and z directions, per unit width of film in the z and x directions
m_{2z}	lower-melt-film downstream mass flow rate per unit width of film
m_{3z}	side melt film downstream mass flow rate per unit width of film
N	screw speed
n	power-law index
n	direction normal to a flow boundary (section 4.4)
n'	gradient of shear-stress curve plotted logarithmically against apparent shear rate
P	pressure drop over rheometer capillary (section 3.3.3)
P	pressure just after the gate to a mould cavity (section 7.3.3)
Pe	Peclet number
P_r	pressure gradient in the radial direction
P_x	pressure gradient in the x direction
P_{x1}	pressure gradient in the x direction at channel inlet
P_z	pressure gradient in the z direction
P_{z1}	pressure gradient in the z direction at channel inlet
P_0	pressure drop for a zero-length capillary
p	pressure
p	compressive stress in a solid plug, acting in the downstream direction (section 6.2)
p_0	downstream pressure at the beginning of a feed section
p_1, p_2	pressures at flow-channel inlet and outlet
ΔP	pressure difference across an injection nozzle
Q	volumetric flow rate (per unit width in the case of an infinitely wide channel, per screw channel in the case of an extruder)
Q'	volumetric flow rate at a particular position along the manifold of a flat film die
Q_F	volumetric downstream flow rate in the clearance

Q_L volumetric leakage flow rate over the flights

Q_s, Q_x, Q_y volumetric flow rates in the s, x and y directions, per unit width of channel

Q_0 reference volumetric flow rate in a mould cavity

Q_1, Q_2, Q_3 components of the volumetric flow rate in the melt pool

q surface heat transfer rate (section 4.4)

q volumetric flow rate per unit width of a flat film die (section 5.3.1)

q volumetric rate of melt influx per unit surface area (section 5.5.2.)

R radius of calender rolls (section 5.4)

R radial distance of flow front from the gate (section 7.3.3)

Re Reynolds number

R_0 reference flow front radius in a mould cavity

R_1 outer radius of disc mould cavity

r radial coordinate in cylindrical polar system

$\bar{r}$ mean distance of a flow channel from the axis of symmetry

r_0 reference radius in a mould cavity

r_1, r_2 inner and outer radii of an annular flow channel

S swelling ratio (section 3.4)

S striation thickness (section 4.7)

S dimensionless shear stress (sections 5.5.1 and 6.3.3)

S screw channel depth ratio (section 6.1.4)

S' striation thickness modified by mixing

s coordinate in the resultant direction of flow

T temperature

T^* dimensionless temperature

$\bar{T}$ bulk mean temperature

T_b temperature of a flow boundary, including extruder barrel

T_m melting-point temperature

T_s temperature of the screw surface

T_0 reference temperature for viscosity measurements

$T_0{}^*$ a dimensionless temperature introduced in equation 5.98

T_1 temperature at flow inlet

$\bar{T}_1, \bar{T}_2$ bulk mean temperatures in the upper and lower melt films

$\bar{T}_{sb}$ bulk mean temperature of the solid bed

T_∞ temperature remote from a flow boundary

ΔT mean temperature rise in a flow

t time

t_{ij} total-stress tensor

t_m local memory time

$t_m{}^*$ dimensionless local memory time

U boundary velocity in the x direction (section 3.2)

U dimensionless velocity in the x direction

U^* dimensionless velocity in the x direction (section 6.1.4)

$\bar{U}$ mean or characteristic velocity in the x direction

U_1 dimensionless x direction velocity at flow channel inlet (section 4.5)

U_1, U_2	peripheral speeds of calender rolls
u	velocity component in the x or r direction
V	velocity of capillary rheometer piston (section 3.3.3)
V	dimensionless velocity in the y direction (section 4.5)
V	wire speed (section 5.1.4)
V	barrel velocity relative to the screw (chapter 6)
V_r	resultant velocity of barrel relative to the solid bed
V_s	resultant relative boundary velocity
V_{sz}	downstream velocity of the solid bed relative to the screw
V_x	boundary velocity in the x direction
V_z	boundary velocity in the z direction
V_1	dimensionless y-direction velocity at flow-channel inlet (section 4.5)
V_1, V_2	relative velocities in the plug-flow analysis (section 6.2.1)
v	velocity component in the y or θ direction
v_i	velocity component in the general x_i direction
W	load applied to drive cone-and-plate rheometer (section 3.3.1)
W	over-all width of a flat film die (section 5.3.1)
W	width of a screw channel (chapter 6)
W^*	dimensionless velocity in the z direction (section 6.1.4)
W_m	width of the melt pool
w	velocity component in z direction
X	dimensionless x direction coordinate
X	width of the solid bed (section 6.3)
x	cartesian coordinate
x_i	general coordinates
Y	dimensionless y direction coordinate
Y^*	dimensionless y direction coordinate (section 6.1.4)
y	cartesian coordinate
y'	cartesian coordinate measured from stress neutral surface (section 6.1.3)
y_0	position of stress neutral point (section 5.4.2)
Z	helical length of a screw channel
Z^*	dimensionless z direction coordinate
z	cartesian coordinate
z_m	helical length of a screw channel required for melting
α	angle of cone-and-plate-rheometer cone (section 3.3.1)
α	pressure coefficient of viscosity at constant shear rate
α	angle of rotation of striation produced by mixing (section 4.7)
α	semiangle of taper of flat film die lips (section 5.3.1)
α	feed angle for solid plug in a screw channel (section 6.2)
β	angle of inclination of flat film die manifold arms
γ	shear rate
$\bar{\gamma}$	mean shear rate
$\gamma_a{}'$	apparent shear rate at the wall of a capillary
$\gamma_t{}'$	true shear rate at the wall of a capillary
γ_0	reference shear rate for viscosity measurements

δ_{ij}	Kronecker delta
$\delta_1, \delta_2, \delta_3$	thicknesses of the melt films at the barrel, screw root and flight surfaces
ϵ_1	initial elastic strain in a cone-and-plate rheometer
ϵ_2	recovered elastic strain in a cone-and-plate rheometer
η_1	generalised viscosity in stokesian constitutive equation
η_2	cross viscosity in stokesian constitutive equation
θ	angle of rotation of cone-and-plate-rheometer cone (section 3.3.1)
θ	angular coordinate in cylindrical polar system
θ	an angle introduced in equation 5.77 (section 5.4)
θ	helix angle of an extruder screw
$\bar{\theta}$	helix angle at the mean channel depth
θ_b	helix angle at the barrel surface
θ_s	helix angle at the screw root
θ_1	an angle introduced in equation 5.77
λ	Trouton extensional viscosity (section 4.3.4)
λ	latent heat of fusion
λ	parameter defining the position of the stress neutral surface (section 6.1.3)
λ	pressure profile parameter introduced in equation 6.81 (section 6.2.1)
λ_1, λ_2	pressure profile parameters introduced in equation 6.87
μ	shear viscosity
$\bar{\mu}$	mean viscosity
μ'	viscosity in the clearance between screw flight and barrel
μ_b, μ_f, μ_s	coefficients of friction at the barrel, flight sides and screw root surfaces
μ_0	reference viscosity for viscosity measurements
π_E	dimensionless channel power consumption
π_P	dimensionless pressure gradient
π_Q	dimensionless volumetric flow rate
π_{Q1}, π_{Q2}	dimensionless volumetric flow rates in the upper and lower melt films
π_X	dimensionless pressure gradient in the transverse channel direction
ρ	density (of melt)
ρ_m	density of a melt
ρ_s	density of a solid polymer
τ	shear stress
$\bar{\tau}$	mean shear stress
τ'	shear stress at the wall of a capillary
τ_{ij}	viscous-stress tensor
$\tau_{xx}^{*}, \tau_{xy}^{*}$	dimensionless stress components
τ_{xy}, τ_{zy}	screw channel shear-stress components evaluated at the barrel surface
$\bar{\tau}_0$	mean shear stress at ambient pressure
τ_1	resultant shear stress in the upper melt film
$\bar{\tau}_1, \bar{\tau}_2$	mean shear stresses in the upper and lower melt films

$\tau_{1z}, \tau_{2z}, \tau_{3z}$	downstream shear-stress components in the upper and lower melt films
φ	a function introduced in equation 5.77 (section 5.4.1)
ϕ	a function introduced in equation 5.99 (section 5.5.1)
ϕ	dimensionless pressure-dependence parameter
ψ	stream function
Ω	rate of rotation of cone-and-plate-rheometer cone
ω	rate of solidification per unit area (section 7.3.4)
ω_{ij}	rate-of-rotation tensor
$\omega_1, \omega_2, \omega_3$	rates of melting per unit area from the top, bottom and side of the solid bed

1

Introduction

The last few decades have seen the rapid development of synthetic polymeric materials to the point where, in a number of countries, their total rate of production exceeds that of metals on a volume basis, and will in due course do so in terms of weight. Along with the expansion in the manufacture of polymers has come the development and improvement of processes to convert them into useful products.

1.1 Polymeric Materials

Polymeric materials, for present purposes, are those synthetic high molecular weight materials that are of commercial importance. They include plastics – of both the *thermoplastic* and *thermosetting* type – and *elastomers* (rubbers). The most important of these are thermoplastics, and the main emphasis of this book is on the processing of such materials, although at least some of the principles discussed can be applied to thermosets and rubbers.

As engineering materials, polymers compare unfavourably with many older materials, notably metals. They lack strength and stiffness, show time-dependent behaviour, are often unexpectedly brittle and can be used only over limited temperature ranges. Their one main advantage, however, which frequently overrides other considerations, is the ease and cheapness with which they can be processed. A single operation can often be used to produce a finished article of considerable geometric complexity, but of high dimensional accuracy and surface finish. The cost of processing rarely exceeds that of the raw material. The mechanical properties of polymers may be improved by adding reinforcing fillers or by making composite materials. Although a wide range of both organic and inorganic fillers are used in thermosets, this is much less commonly done with thermoplastics, for the very good reason that the resulting mixtures are less easy to process.

The most important thermoplastics, in terms of rates of production, are low- and high-density polyethylenes, polypropylene, polystyrene, polyvinyl chloride and nylons. All of these are available in many grades, having properties appropriate for different applications and processing techniques. Differences between grades are often due to differences in mean *molecular weight* (reflecting the average size of the long-chain molecules) and *molecular weight distribution* (reflecting the variation of molecule size about the mean). In addition to reinforcing fillers and colourants, various additives may be used to affect the properties of the material either during processing or in later use. These include *lubricants* and *plasticisers* to facilitate processing, stabilisers against *degradation* (breakdown of the molecules) by heat or light, and *fire retardants* to make the product less inflammable.

Thermoplastics, as their name suggests, melt reversibly on heating and are usually shaped into the required form while in the molten or *melt* state. Although melts can best be described as fluids, they behave very differently from most familiar fluids. For example, they display significant elastic properties, and should therefore be regarded as *viscoelastic*. Also, melt viscosities are both very high and non-newtonian.

1.2 Polymer Processing

Polymer processing is concerned with the operations carried out on polymeric materials to form them into useful products. The chemical processes involved in the manufacture of polymers from their monomers are specifically excluded. In the case of thermoplastics, the main steps in any process are first to melt, then to shape and finally to cool the material in its new form. The heat required for melting may be supplied by radiation or conduction, or by mechanical work. Mixing of the melt is desirable to improve the properties of the product.

Thermoplastics are normally obtained from the polymerisation reactor in the form of either powder or melt. While it is possible for a processor to purchase some materials in powder form for direct conversion into a finished product, the polymer manufacturer often carries out an *homogenising* or *compounding* operation after polymerisation. Homogenisation serves to mix the raw polymer thoroughly, in order to break up lumps of high-molecular-weight material, for example, and also to remove unconverted monomer or other volatiles. It also permits the inclusion of additives, and sale of the material in the form of *granules*, which are easier to use than powder in the subsequent product-processing operations. Homogenisation is carried out in very large *screw extruders* or other continuous mixing equipment. Screw extruders and product processing equipment are discussed in chapter 2.

Two main methods are used to make polymer granules. With a technique known as underwater die face cutting, the melt is forced through a multihole die into water, where the emerging strands are cut at the die face to form discrete particles; the particles are roughly spherical in shape when solidified. Alternatively, extruded strands are solidified by cooling in a water bath before being cut; this method, known as lace cutting, gives granules of cylindrical shape. The term granulation in the present context of polymer manufacture should not be confused with the methods used to reduce the size of process scrap prior to reprocessing.

1.3 Analysis of Polymer Processes

Although polymer processes have often been evolved by largely trial-and-error methods, the performances currently being demanded make it increasingly important not only to understand how such processes function, but also to be able to make reliable quantitative predictions of performance. Provided the mathematical analyses used to do this embrace the important features of the material behaviour under processing conditions, it is possible to consider how to improve both the performance of existing machines and the design of new ones. Even if a full analysis of a process is not possible or practical, an appreciation of what are the important features is often of considerable assistance in solving processing problems.

The first attempts to analyse polymer processes used only very simple models of processes and material behaviour. For example, melt viscosity was assumed to be constant, and heat-transfer effects in flowing polymers were virtually ignored. The agreement with observed phenomena was poor. In recent years, however, the need for more sophisticated models has been recognised, and with the advent of high-speed digital computers it is now possible to solve the resulting sets of mathematical equations economically.

1.4 Scope of the Book

The main objective of this book is to promote an understanding of the principles of at least some of the more important polymer processes. To this end, analyses of these processes are developed, attempting to take into account all those physical phenomena that appear to have a significant influence on process performance.

In view of the number of processes in use, it is not possible to treat them all in depth in a book of the present size. A number of somewhat arbitrary decisions have therefore been made on what is to be included, and these inevitably reflect to some extent the author's interests and experience. As already indicated, attention is largely confined to thermoplastic materials. Similarly, the emphasis is on flow behaviour inside processing machinery, and on the design of internal flow channels. The mechanical design of the machine components for strength, stiffness and resistance to wear is not considered explicitly. Relatively little attention is given to process stability and control, on the basis that steady operation needs to be thoroughly understood before unsteady behaviour can be examined. Also, no detailed consideration is given to free surface flows of polymer films and filaments. Consequently, it is not necessary to give more than superficial attention to the elastic properties of melts.

After a brief review of the main product processes in chapter 2, chapter 3 is concerned with polymer properties relevant to processing. Chapter 4 establishes the mathematical fundamentals of melt flow, while the remaining chapters concentrate on the analysis of processes, particularly screw extrusion and injection moulding. For more detailed descriptions and discussion of practical equipment, and the sources of some of the analyses presented, the reader is referred to the list of further reading.

2

Introduction to the Main Polymer Processes

The principal methods used to process thermoplastic materials into finished or semifinished products are, in order of importance: screw extrusion, injection moulding, blow moulding and calendering. The reason for the outstanding importance of extrusion is that almost all thermoplastics pass through an extruder at least once during their lives, if not during fabrication of the end product, then during the homogenisation stage. Also, in most modern injection moulding machines, the preparation of melt for injection is carried out in what is essentially a screw extruder. Similarly, blow moulding and calendering are often post-extrusion operations. An important distinction exists between extrusion and calendering on the one hand, and moulding techniques on the other, in that while the former are continuous processes, the latter are discontinuous.

Processing methods for thermosetting materials, which are irreversibly polymerised during processing, are largely confined to simple moulding techniques. Rubbers, on the other hand, can be both extruded and moulded.

2.1 Screw Extrusion

The extrusion process is one of shaping a molten polymeric material by forcing it through a *die*. By far the most common method of generating the required pressure, and usually of melting the material as well, is by means of one or more screws rotating inside a heated barrel. While the form of the die determines the initial shape of the *extrudate*, its dimensions may be further modified — for example, by stretching — before final cooling and solidification takes place.

2.1.1 Single-screw Extrusion

Figure 2.1 shows the diagrammatic cross-section of a typical single-screw *plasticating* extruder, which is required to melt and to pump the polymer. Solid material in the form of either granules or powder is usually gravity fed through the

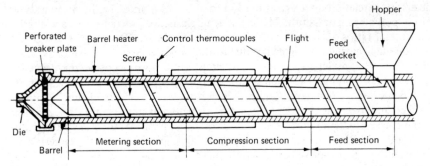

Figure 2.1   *Diagrammatic cross-section of a typical single-screw plasticating extruder*

hopper, although crammer-feeding devices are sometimes used to increase feed rates. After entering a helical screw channel via the feed pocket, the polymer passes in turn through the *feed, compression* and *metering* sections of the screw. The channel is relatively deep in the feed section, the main functions of which are to convey and compact the solids. The compression section owes its name to its progressively decreasing channel depth. Melting occurs there as a result of the supply of heat from the barrel and mechanical work from the rotation of the screw. The shallow metering section, which is of constant depth, is intended to control the output of the machine, generate the necessary delivery pressure and mix the melt. On leaving the screw, the melt is usually forced through a perforated *breaker plate*. A renewable *screen pack*, consisting of layers of metal gauze, may also be used; like the breaker plate, this serves to hold back unmelted polymer, metal particles or other foreign matter, and to even out temperature variations. Single-screw extruders are essentially screw viscosity pumps, which rely on material adhering to the barrel for their conveying and pumping action. Thanks to high melt viscosities, pressures of 40 MN/m^2 or more can be achieved at the delivery end of the screw.

The screw is held in position by an axial thrust bearing – not shown in figure 2.1 – and driven by an electric motor via a reduction gearbox. Screw speeds are generally within the range 50 to 150 rev/min, and it is usually possible to vary the speed of a particular machine over at least part of this range. The radial clearance between the *flights* of the screw and the barrel is small, and the surfaces of both are hardened to reduce wear.

Barrel and die temperatures are maintained by externally mounted heaters, typically of the electrical-resistance type. Individual heaters or groups of heaters are controlled independently via thermocouples sensing the metal temperatures, and different *zones* of the barrel and die are often controlled at different temperatures. The region of the barrel around the feed pocket is usually water cooled to prevent fusion of the polymer *feedstock* before it enters the screw channels. Cooling may also be applied to part or all of the screw by passing water or other coolant through a passage at its centre, access being via a rotary union on the driven end of the screw.

The size of an extruder is defined by the nominal internal diameter of the barrel. Sizes range from about 25 mm for a laboratory machine, through 60–150 mm for most commercial product extrusions, up to 300 mm or more for homogenisation during polymer manufacture. Modern thermoplastic extruders have screw length-to-diameter ratios of the order of 25 or more. Screws are often single start, that is, have one helical flight, with leads equal to their nominal external diameters. The channel depths and lengths of the three screw sections vary considerably, according to the type of polymer and application. An important characteristic of a screw is its *compression ratio*, one definition for which is the ratio between channel depths in the feed and metering sections. This ratio normally lies in the range 2—4, according to the type of material processed. Output rates obtainable from an extruder vary from about 10 kg/h for the smallest up to 5000 kg/h or more for the largest homogenisers. Screw-drive power requirements are usually of the order of 0.1—0.2 kW h/kg.

Many modifications to the basic form of screw design are used, often with the aim of improving mixing. Another variant is the *two-stage* screw, which is effectively two screws in series. The decompression of the melt at the end of the first stage, where the screw channel suddenly deepens, makes it possible to extract through a vent any air or volatiles trapped in the polymer.

2.1.2 Multiscrew Extrusion

Extruders having more than one screw are also used for some applications, twin-screw machines being the most common. Such extruders can have two screws intermeshing or not quite intermeshing, corotating or counterrotating. The more common intermeshing type have distinct advantages over single-screw machines in terms of an improved mixing action, and are not so much screw viscosity pumps as positive displacement pumps. They are, however, considerably more expensive.

2.1.3 Extrusion Dies

The simplest extrusion dies are those used to make axisymmetric products such as lace and rod. The main design consideration with such dies is that changes in flow channel diameter from that of the extruder barrel bore to that of the die exit are gradual. Smooth melt flow is thus ensured, with no regions where material can be retained and degrade. In designing dies for more complicated profiles, due allowance must also be made for elastic recovery, which may cause changes in shape after the extrudates leave the dies. Other types of extrusion die are used in the production of flat film, sheet, pipe and tubular film, and in covering wire and cable.

2.1.4 Flat-film and Sheet Extrusion

The distinction between flat film and sheet is one of thickness, both being extruded in similar types of dies. As the widths of such flat sections are much greater than the extruder-barrel diameters, the dies must spread the melt flow laterally and produce extrudates of as uniform a thickness as possible. Figure 2.2 shows a diagrammatic view of a die meeting these requirements, which is of the *coat-hanger* type, so

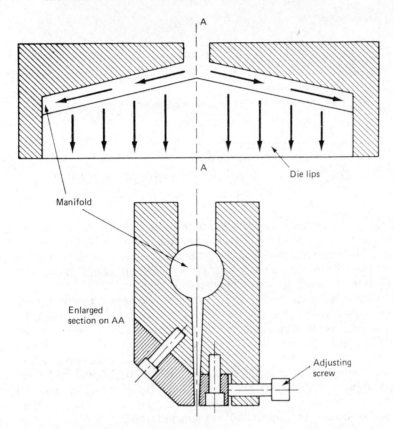

Figure 2.2   *Coat-hanger die for flat film or sheet*

named after the shape of the manifold that distributes the melt to the die lips. By adopting this shape, the length of the narrow tapered channel leading up to the die exit is varied in such a way as to equalise the flow resistance along any path taken by the melt, thereby promoting uniformity of extrudate thickness. Some final die lip adjustment is possible to overcome minor thickness variations.

Cooling of extruded flat film or sheet can be achieved either by quenching in water or, in the case of film, by wrapping around a water-cooled roll. The latter is known as *chill roll casting*. In a variation of this process, known as *extrusion coating*, the hot melt film is sandwiched under pressure between the chill roll and, say, a layer of paper, to which it adheres to give a laminated product.

2.1.5 Pipe Extrusion

Figure 2.3 shows a die, sizing and cooling arrangement for producing extruded pipe. The die is of the required annular shape, the *torpedo* that forms the inner surface of

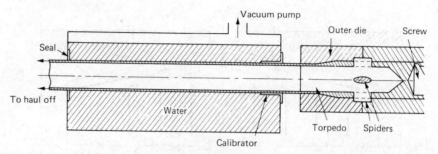

Figure 2.3   *Production of extruded pipe using a vacuum cooling bath*

the annulus being supported by three or four *spiders*. Despite streamlining of these narrow obstructions, they tend to cause both circumferential wall thickness variations and longitudinal weaknesses in the pipe. Some adjustment of the outer die relative to the torpedo is possible to correct pipe wall eccentricity.

After leaving the die, the extrudate passes through a calibrator, which imparts the required external diameter, into a water bath maintained at below ambient pressure to prevent collapse of the pipe before it has cooled and stiffened. The necessary axial motion of the pipe is maintained by a *haul off* device (not shown in figure 2.3) which grips the outside of the pipe. Other types of pipe die and cooling arrangements are also used. For example, instead of a vacuum cooling bath to maintain the required pressure difference, the free end of the pipe can be sealed and internal air pressure applied via a radial hole along one of the spiders emerging at the die face in the centre of the torpedo. Also, crosshead dies similar to those described in the next two subsections are sometimes used.

2.1.6 Tubular Film Extrusion

The most common method of producing thermoplastic film is via the tubular film blowing process, a general view of which is shown in figure 2.4a. Melt from the extruder is turned through a right angle and extruded vertically upwards in the form of a thin tube. This tube is expanded by internal air pressure to form a bubble, and stretched in the direction of flow by means of powered nip rollers, which also close the bubble. The flattened film is then wound on to a reel. Cooling of the melt once it has left the die is accomplished by a flow of air from a cooling ring directed on to the bubble. Solidification occurs at a *freeze line*, which is often observable near where the maximum bubble diameter is first reached.

Figure 2.4b shows a cross-section of one type of tubular film die, known as a side-fed or *crosshead* type because the direction of flow is turned through a right angle. The flow is obliged to divide around the central torpedo, which contains the bubble inflation air inlet, recombining before the die lips. After leaving the die, the melt flow is of the free-surface type; the bubble size, shape and final mechanical properties depend on a very complex interplay of melt viscosity, elasticity, surface tension and heat transfer.

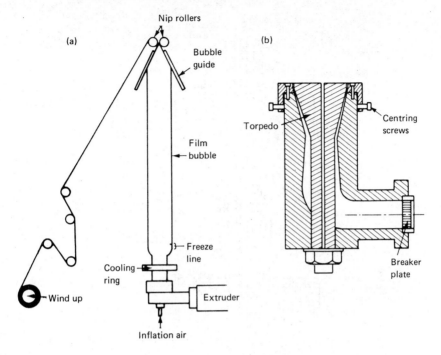

Figure 2.4   *The tubular film-blowing process: (a) general view of the process; (b) cross section of a crosshead tubular film die*

2.1.7 Wire and Cable Covering

Wire and cable covering operations are carried out over a very wide range of line speeds, from about 1 m/min for large high-voltage electrical cables to 1000 m/min or more for small-diameter wires. Nevertheless, the types of die used are similar, being of the crosshead type to accommodate the conductor entering at an angle — often a right angle — to the axis of the extruder. Figure 2.5 shows a typical die for applying a single layer of polymer to a conductor drawn at constant speed through its centre. The success of such an arrangement depends on the design of the flow deflector, which serves to distribute the melt into a layer of uniform thickness on the conductor. A further view of a typical deflector is shown in figure 5.7.

2.2 Injection Moulding

In the injection-moulding process, molten polymer is forced under high pressure into a closed mould of the required shape, where it is cooled before the mould is opened and the finished article extracted. Early injection moulding machines melted and injected material by means of a heated cylinder and plunger arrangement. For reasons of improved melt quality and faster production rates, this simple

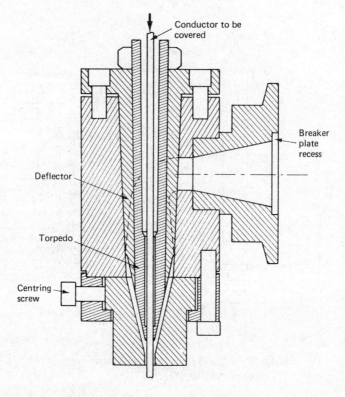

Figure 2.5   *Cross section of a wire or cable covering crosshead*

plunger type of machine has been almost entirely superseded by the single-screw —
or occasionally twin-screw — type, which employs what is essentially a screw
extruder to both melt and inject the polymer.

2.2.1 Screw Injection-moulding Machines

Figure 2.6 shows a simplified diagrammatic cross-sectional view of a typical
horizontal single-screw injection moulding machine. Solid polymer in either granule
or powder form supplied from a hopper is processed by the screw into a melt, which
accumulates between the end of the screw and the injection *nozzle* at the end of the
barrel. The screw, which is free to move axially, is forced back until sufficient melt
to fill the mould has accumulated, when screw rotation is stopped. Injection is
achieved by forward displacement of the screw as a result of an hydraulically
applied axial load on its rear end. The high pressures involved necessitate the use of
a non-return mechanism in the form of a ring valve on the tip of the screw, in order
to prevent backflow along the screw channel.

Melt entering the mould flows along a system of passages to a cavity, or cavities,

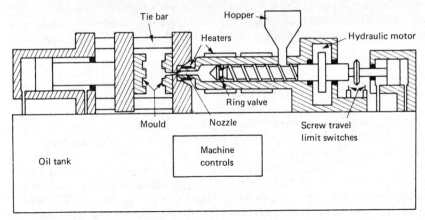

Figure 2.6   *Single-screw injection-moulding machine*

of the required shape, where it solidifies in contact with the cold mould surfaces. During injection, the mould is held closed by an hydraulic or mechanically operated clamping system. The maximum available clamping force influences both the maximum area of mould cavities normal to the direction of clamping, and the maximum permissible injection pressure. Machine capacities are specified in terms of both mould clamping force and maximum weight of material per injection or *shot*. While the former ranges from a few tonnes to several thousand tonnes, the latter can vary from a few grams to tens of kilograms. Maximum injection pressures are generally in the range $80-200$ MN/m^2.

Screws used in injection-moulding machines are shorter than those used in extruders, and usually have length-to-diameter ratios of between 15 and 20. The effective lengths are even less, because feed sections must be long if the screws are to be able to operate in their fully back positions. Melt quality is therefore relatively poor, but this is largely overcome by the additional mixing imparted when the material is forced through the nozzle and mould passages.

2.2.2 The Moulding Cycle

A typical moulding cycle may be outlined as follows. The mould is first closed, and the screw moves forward to inject melt. The screw is held in the forward position under pressure, during which time the melt in the mould cools and contracts, allowing a little more material to enter. Once the *gates* at the entries to the mould cavities have frozen, the screw may be withdrawn. Meanwhile, further thermal contraction of the mouldings is to at least some extent offset by the expansion due to relaxing pressure in the cavities.

When the gates have frozen, the whole screw-injection unit is normally retracted from the mould, both to break the polymer thread at the nozzle exit and to prevent excessive cooling of the nozzle in contact with the cold mould. Screw rotation then takes place to provide a new charge of melt. When sufficient cooling time has been allowed for the mouldings to become stiff enough to be ejected, the mould opens, usually with the mouldings clinging to the mould face remote from the injection point. Ejector devices are then used to detach the mouldings, which fall clear under gravity.

2.2.3 The Mould

Figure 2.7 shows the layout of a typical simple mould, which has two identical cavities. Melt from the nozzle enters via the *sprue*, which has a divergent taper to facilitate removal when frozen. Opposite the sprue is a *cold slug well*, which serves both to accept the first relatively cold portion of injected material, and to allow a re-entrant shape on the end of an ejector pin to grip the sprue when the mould opens. The remaining melt flows along a system of *runners* leading to the mould cavities. The gates at the entries to the cavities are very narrow in at least one direction, so that the mouldings will be readily detachable from the runners after removal from the mould. When the two halves of the mould shown in figure 2.7 part, the mouldings, runners and sprue are obliged to remain with the left-hand half. Final removal is then effected by movement of the ejector pin from left to right, which releases the trapped cold slug well. In larger and more complicated moulds, more sophisticated systems of ejector pins and other devices are used. Also, multiplate moulds and more elaborate runner systems may be necessary.

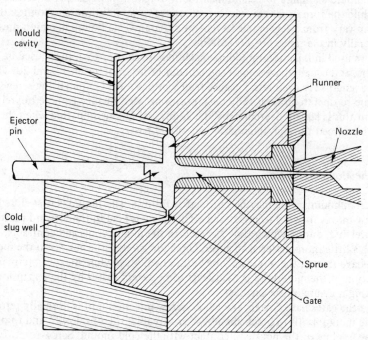

Figure 2.7   *General layout of a mould*

Some general comments on the mechanism of flow in a mould cavity are appropriate. As melt comes into contact with the cold metal, it freezes to give a thin layer of solid. Low polymer thermal conductivity ensures that the growth of this layer is slow. At moderate rates of injection, the flow spreads outward in all available directions away from the gate, with a core of hot melt flowing between solidified skins. At much higher injection rates, however, the melt 'jets' through the gate and across the mould cavity, filling the far side first in a very irregular manner. This filling mechanism is to be avoided, because the properties of the finished moulding are inferior.

The sandwich moulding process takes full advantage of the skin formation in the normal mechanism of mould flow. Two different polymers are injected, one after the other, through the same gates, the first one forming the outer skin of the finished article, and the second one its core. Thus, a cheap weak core can be injected inside a skin of higher-quality material. A remarkable uniformity of skin thickness can be achieved in practice.

2.3 Blow Moulding

Hollow articles, such as bottles and other containers, can be manufactured by the blow-moulding process. The most common procedure is to extrude vertically downwards, from a crosshead die, a hollow thick walled tube or *parison* of melt. This tube is surrounded by a split mould of the appropriate shape, one end being clamped around a spigot incorporating an inlet for compressed air, which is used to blow the parison into contact with the water-cooled mould. When solidified, the moulding is removed and its ends are trimmed of excess material.

While blowing and cooling are being carried out, the mould unit, of which there may be several associated with one extruder, is removed from beneath the extrusion crosshead to allow a new parison to form. Figure 2.8 shows a simplified view of this process. In figure 2.8*a*, the two halves of the mould are about to close on the parison, thereby severing it at the top and squeezing it into contact with the spigot at the bottom. Figure 2.8*b* shows the parison blown into contact with the mould.

Both the elastic properties of the melt and the effect of gravity on the suspended parison affect its thickness distribution, and only relatively modest dimensional accuracy can be achieved. Also, the amount by which the diameter of a parison can be increased by blowing is limited. In addition to extrusion blow moulding, there is also an injection blow-moulding process, in which the material to be blown is first injection moulded. Although better dimensional accuracy than in extrusion blow moulding can be achieved, the process is more expensive.

2.4 Calendering

Calendering is a process for producing continuous sheet by squeezing molten polymer between rotating horizontal rolls. It is most commonly applied to poly-vinyl chloride, and is capable of maintaining very close dimensional tolerances, as well as imparting a textured surface to the finished sheet. A calender unit consists of several rolls, typically four arranged as shown in figure 2.9; the speeds and

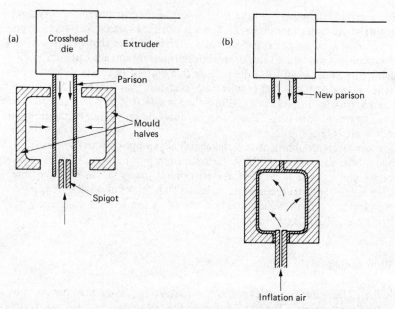

Figure 2.8   *The extrusion blow-moulding process: (a) parison about to be enclosed in the mould; (b) severed parison blown into contact with the mould*

temperatures of the rolls are independently controlled. The minimum distances or *nips* between successive pairs of rolls are made progressively smaller. Molten material, often supplied by a screw extruder, is fed to the calender in the form of a continuous strip. At each nip, it is sheared, mixed and reduced in thickness. The material is encouraged to take the intended path shown through the calender by running successive rolls at progressively higher speeds and temperatures: the melt

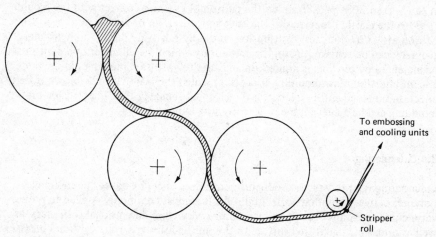

Figure 2.9   *Typical arrangement of a four-roll calender*

tends to adhere to the faster and hotter roll of a pair. Sheet is removed from the last calender roll by a small high-speed stripper roll, which may impart some stretching. Special surface finishes may then be applied by embossing rolls before the sheet is cooled in contact with a series of chilled drums.

2.5 Other Processes

There are several other polymer processes of relatively minor importance in terms of volume of production. For example, *thermoforming* is a sheet manipulation process in which a sheet of thermoplastic material is first heated to its softening temperature and then forced by the application of either vacuum, positive air pressure or a forming tool to take up the shape of a cool mould.

Another process that is very important for the production of fibres and filaments is that of *spinning*. Melt supplied by either an extruder or gear pump is forced vertically downwards through a series of very small holes in a flat plate or *spinneret*, and the resulting threads are air cooled and rapidly stretched by winding at high speed on to a bobbin.

Many thermoplastics can be worked by engineering techniques including welding, cutting and machining, although to do so to any significant extent is to lose the advantage offered by polymeric materials over metals in terms of ease of fabrication.

2.6 Effects of Processing

Any processing operation applied to a polymer may have both chemical and physical effects on the material, some beneficial and some harmful to the properties of the finished product. Taking chemical effects first, *degradation* – breaking of the long-chain polymer molecules into a larger number of smaller molecules – can occur in different ways. Subjection of the material to high shear stresses tends to break the longer molecules preferentially. Thus, shear degradation affects not only the molecular weight, but also the molecular weight distribution. Thermal degradation, on the other hand, which becomes progressively more severe as the processing temperature increases, can affect molecules of any length. The same is true of oxidative degradation, which occurs in the presence of oxygen at elevated temperatures.

Physical changes are not necessarily undesirable. For example, the effect of deformation during melt flow is to promote intimate mixing of the material and its additives. Such deformation also causes molecular *orientation*, with the molecules tending to disentangle and to take up positions lying along streamlines in the flow. If this orientation is retained by subsequent cooling, the resulting product has anisotropic properties, enhanced in the direction of orientation but inferior in directions normal to this. For example, the high degree of axial orientation imparted during fibre spinning helps to give synthetic fibres their outstanding properties. Also, extruded tubular film is biaxially oriented, in the axial and circumferential directions of the bubble. Other physical effects of processing include loss of additives by evaporation, migration of additives to the flow boundaries, absorption of water during cooling operations and residual stresses set up by thermal contraction.

3

Processing Properties of Polymers

In this chapter, the more important physical properties of polymeric material which affect their behaviour in processing operations are discussed. The purpose is to provide an introduction to flow and heat transfer characteristics, rather than an exhaustive treatment of polymer properties. A list of further reading on the subject is provided at the end of the book.

One of the limitations to the methods of analysing processing techniques described in subsequent chapters is often the lack of adequate property data. Considerable effort is needed to obtain these data and they must be revised as polymer properties change with variations in the manufacturing processes, and as new grades replace old ones. In some cases, there are significant property variations from batch to batch. Single values of particular properties are often inadequate. Melt viscosities, for example, depend on the rate at which the material is deformed, and many properties are affected by both temperature and pressure.

3.1 Melting and Thermal Properties of Polymers

Unlike conventional materials that exhibit a distinct phase change from solid to liquid at a particular temperature, the melting point, polymers undergo an equivalent change over a finite temperature range. Below this range, their behaviour is that of elastic solids showing some viscoelasticity in the form of time-dependent properties. Above the melting range, polymers in the melt form may be regarded as highly viscous liquids exhibiting some elastic effects.

Figure 3.1 shows a typical plot of specific enthalpy against temperature for a polymeric material. The melting temperature range, which is typically of the order of 10 °C, normally occurs between about 100 °C and 300 °C, depending on the material concerned. For processing, the minimum satisfactory temperature is generally some 10–20 °C above the upper limit of this melting range. Maximum processing temperatures may well be imposed by susceptibility to thermal degradation. The portions of the enthalpy curve in the solid and melt regions are not normally either straight lines or parallel to each other. In other words, the

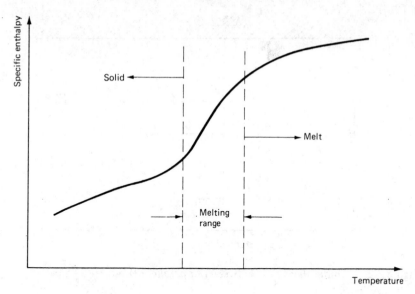

Figure 3.1   *Typical enthalpy–temperature curve for a polymer*

specific heats of solid and melt, which are given by the gradients of the relevant portions of the curve, are not equal and may be significantly temperature dependent. For the purposes of process analysis over comparatively narrow ranges of temperature, however, such temperature dependence is often neglected.

Within the melting temperature range the change of enthalpy is smooth compared to the abrupt increase at the melting point associated with the latent heat of fusion of conventional materials. Nevertheless, it is sometimes expedient for the purposes of analysis to express the change in terms of a latent heat at a distinct melting point. The other thermal property which is of very considerable importance in processing operations is the thermal conductivity, of both solid and melt. Because polymer conductivities are low, typically of the order of one per cent of that of steel, they are difficult to measure accurately. While they should ideally be regarded as functions of temperature, pressure and rate of deformation, very few attempts have been made to determine the forms of dependence.

3.2 Viscous Properties of Polymer Melts

In addition to being very viscous under processing conditions — typically at least a thousand times more viscous than, for example, lubricating oils — polymer melts are also non-newtonian, that is, their viscosities depend not only on temperature and pressure, but also on the rate at which the material is deformed. In order to discuss the practical implications of this effect, it is appropriate to review the definition of viscosity.

Consider the simple shear flow between flat parallel plates illustrated in figure 3.2. The upper plate moves at a velocity U relative to the lower one in the direction

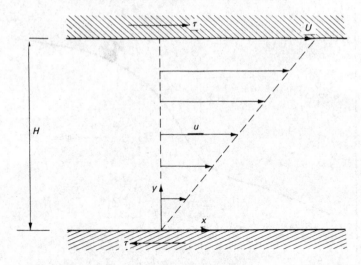

Figure 3.2   *Simple shear flow between flat parallel plates*

of the x coordinate; coordinate y is normal to both plates. The velocity of the flow, u, is proportional to the distance y above the stationary plate, and the constant shear rate is

$$\gamma = \frac{du}{dy} = \frac{U}{H} \tag{3.1}$$

where H is the distance between the plates. A shear stress, τ, applied as shown to each plate, maintains the shearing motion, and viscosity is defined as the ratio of stress to shear rate

$$\mu = \frac{\tau}{\gamma} \tag{3.2}$$

Measurements of polymer melt viscosity over wide ranges of shear rate — obtained, for example, by the methods described in section 3.3 — show results of the form illustrated in figure 3.3. Viscosity and shear rate are plotted on logarithmic scales to form the *flow curve* of the material. When sheared at low rates, typically less than 10^{-1} s^{-1}, melts behave in a newtonian manner. At higher rates of shear, however, particularly over the range 10^{1}–10^{3} s^{-1} relevant to most processing operations, melt viscosity decreases substantially with increasing shear rate. In other words, melts are *pseudoplastic* materials. There are no common polymers that show the reverse trend of increasing viscosity associated with *dilatant* materials.

Despite the changes in slope of the typical flow curve, it may be approximated by a straight line over a reasonably narrow range of shear rate. Since this is on a logarithmic plot, it implies a power-law relationship between viscosity and rate of shear. In practice, it is more often the relationship between shear stress and rate which is represented by a power-law equation of the form

$$\tau = C\gamma^{n} \tag{3.3}$$

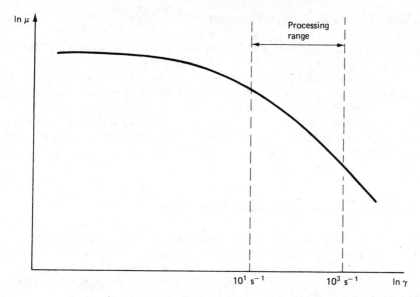

Figure 3.3   *Typical flow curve for a polymer melt*

where C is a melt consistency and n is the power-law index. Using the definition of viscosity given by equation 3.2

$$\mu = \frac{\tau}{\gamma} = C\gamma^{n-1} \tag{3.4}$$

It must be emphasised that equation 3.3 is a purely empirical result obtained by curve-fitting experimental data over a particular range of shear rate. Such a range of interest is typically of the order of one decade. For example, in melt flow in screw-extruder channels, the shear rate range of greatest significance is about $10-10^2$ s^{-1}, while in flows in injection nozzles it is often 10^2-10^3 s^{-1}. Different portions of the flow curve, and somewhat different values of C and n, would therefore be appropriate for these two applications. Within shear rate ranges of one or two decades, the accuracy of the power-law representation of flow data is comparable with that of the experimental data themselves, which is usually better than 5 per cent.

The main advantage of using the empirical power law is that it allows flow data to be represented in terms of just two parameters, C and n. Both, however, should be regarded as functions of temperature and pressure. Typical values of n for some common thermoplastics are: polyethylene 0.3–0.6, polypropylene 0.3–0.4, polyvinyl chloride 0.2–0.5 and nylon 0.6–0.9. All of these are significantly lower than the newtonian value of unity. Although useful for describing experimental data, equations 3.3 and 3.4 are not convenient for mathematical manipulation. The units of C depend on n, and negative shear rates are not admissible. Ways of overcoming these limitations are described in chapter 4.

3.3 Methods of Measuring Melt Viscosities

The commonly used methods of measuring polymer-melt viscosities are of two types namely rotational and capillary. In the rotational type, melt is sheared between rotating surfaces. The rate of shear is derived from the relative speed while the shear stress is obtained from the torque required to maintain the motion. In the capillary type, melt is forced to flow along a small tube. A rate of shear is derived from the steady rate of flow achieved, while the associated shear stress is calculated from the measured pressure gradient along the tube. Several methods are considered in detail, the main purpose being to emphasise their limitations and to indicate sources of error in the resulting viscosity data.

3.3.1 The Cone-and-plate Rheometer

Figure 3.4a shows a diagrammatic view of a simple cone-and-plate rotational rheometer. A nearly flat cone is mounted with its axis normal to a plate, the usual material for both components being steel. The plate is heated to the required temperature and the cone is enclosed to minimise heat losses to the surroundings. The shaft on which the cone is mounted is supported between bearings. It may either be driven at constant speed and the applied torque measured or, as in figure 3.4a, a known torque may be applied by a simple weight, string and pulley arrangement. In this case, the speed of rotation is measured, often with the aid of an angular scale on the cone shaft.

The experimental procedure may be outlined as follows. With the cone raised from the plate, a sample of solid polymer in either granule or powder form is placed on the plate. Sufficient time is then allowed for it to melt and reach the required temperature before the cone is lowered and its tip brought just into contact with the

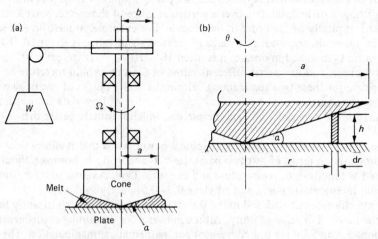

Figure 3.4   *A cone-and-plate rheometer: (a) the instrument; (b) section through the melt flow*

plate. A particular load W is applied to turn the cone, and the steady speed of rotation, Ω, achieved is measured. The geometric parameters of the apparatus which need to be known are the cone angle, α, which normally lies in the range $1-5°$, the radius of the cone, a, and the distance from the cone axis, b, at which the load is applied.

The required shear rate and stress may be derived from a simple analysis of the melt flow. Consider the small annular element of melt shown in figure 3.4b. Its radial width is dr and it lies at a radius r from the cone axis. Provided the cone angle is small, this element is effectively subjected to simple shear flow of the type illustrated in figure 3.2. The relative velocity between the surfaces bounding the element is $r\Omega$, where for steady conditions

$$\Omega = \frac{d\theta}{dt} \tag{3.5}$$

In this expression, t is the time and θ the angular displacement of the cone. Hence, using equation 3.1, the shear rate in the element is

$$\gamma = \frac{r\Omega}{h} = \frac{r\Omega}{r\tan\alpha} = \frac{\Omega}{\tan\alpha} \approx \frac{\Omega}{\alpha} \tag{3.6}$$

where h is the height of the element. Although the approximation of $\tan\alpha$ by α is only valid for small angles, α must be small in order to ensure that the flow can be treated as simple shear. The most important feature of the final expression for shear rate is that it is independent of radius.

If the shear rate is constant throughout the sample, then so is the shear stress, τ. The torque applied to the typical element of melt is the product of this stress, the area of the element in the plane of the plate, $2\pi r\, dr$, and its distance from the axis of rotation, r. Hence, the total torque applied to the cone is

$$\int_0^a 2\pi r^2 \tau\, dr = \frac{2}{3}\pi a^3 \tau \tag{3.7}$$

Now, this torque is equal to the product of the applied load, W, and the radius of application, b, giving the following expression for the shear stress

$$\tau = \frac{3Wb}{2\pi a^3} \tag{3.8}$$

In practice, the value of W used to find τ may need to be corrected for frictional losses found by repeating the experiment in the absence of the polymer. Finally, the viscosity is calculated as the ratio of shear stress to shear rate. Clearly, the procedure can be repeated using different applied loads and temperatures.

The principal advantages of the cone-and-plate rheometer are that it is simple to use and gives stresses and shear rates directly. Because these are constant throughout the flow, the above derivations are equally valid for newtonian and non-newtonian materials. The main disadvantage is the limited range of shear rate that can be covered. A minimum shear rate is reached when the applied torque is of the same order of magnitude as that required to overcome friction in the bearings. Of much greater practical importance, however, is the maximum shear rate, which is limited by the tendency for cavities to form in the melt near the edge of the cone, invalidating the simple shear flow assumption. The range of usefulness of cone-and-plate rheometers is further discussed and compared with other methods of measurement in section 3.3.4. A further minor disadvantage of cone-and-plate instruments is that the same sample is used throughout a test, or even a series of tests. This may result in molecular orientation and thermal degradation, both of which can affect viscous properties.

3.3.2 The Melt Flow Index Test

The *melt flow index* test is of the capillary type and involves the use of a short capillary die at the bottom of a heated barrel, as illustrated diagrammatically in figure 3.5. Solid polymer is loaded into this barrel and, after allowing time for it to reach the required temperature, a piston carrying a simple dead-weight as shown is added. The melt-flow index is derived as the weight of material in grams extruded over a period of 10 min. The dimensions of the apparatus are laid down by BS 2782 and ASTM 1238 and the principal ones are given in the figure caption. Both the test temperature and the size of the weight depend on the material concerned. For example, for polyethylenes probably the most commonly used conditions are 190 °C and a weight of 2.16 kg.

The main advantage of the melt flow index test is that it is a standard one widely used in the plastics industry. It provides a simple measure of flow properties, the merits of which are further discussed in section 3.3.4. The melt flow indices of

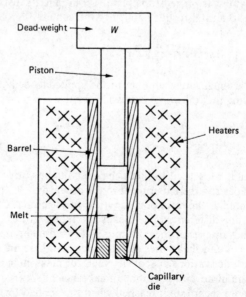

Figure 3.5   *Apparatus for the melt flow index test; standard dimensions are: barrel diameter 9.550 mm, capillary diameter 2.095 mm, capillary length 8.000 mm*

materials vary inversely with their viscosities. Because the capillaries used are so short, however, it is not possible to derive reliable values of viscosity from these indices. The reasons for making this assertion will become clear when the more general type of capillary rheometer is considered.

3.3.3 The Capillary Rheometer

Capillary rheometers, which are sometimes known as ram extruders, can be regarded as refinements of the apparatus for measuring melt flow indices. The capillary is generally much longer and the rates of flow of melt can be varied over wide ranges. Two main types of instrument are used. In the constant-rate type illustrated in figure 3.6a, a rigid piston is driven towards the capillary at a constant velocity, V. Figure 3.6b illustrates the different piston arrangement used in the constant-pressure type where gas, usually nitrogen, is applied under pressure to a small floating piston on top of the melt. In either type, the pressure in the melt is measured either at the piston or by means of a transducer in the barrel wall just before the entrance to the capillary.

The volumetric flow rate, Q, may be obtained in the constant-rate type as the product of piston velocity and barrel cross-sectional area, while in the constant-pressure type it must be measured directly. As it is mass flow rate that can be measured more readily — by weighing a sample collected over a known time interval — the constant-pressure type of instrument requires a knowledge of polymer-melt density

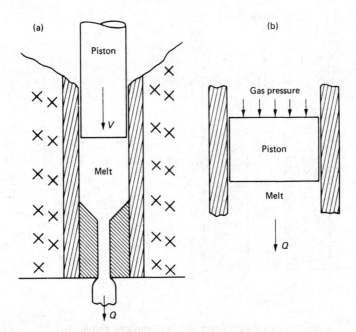

Figure 3.6   *Capillary rheometers: (a) constant-rate type; (b) constant-pressure type*

at the temperature of the test. With the constant-rate type, however, weighing the extrudate provides a useful means of measuring melt density in addition to viscosity. Other advantages of this type are that pressures up to ten times those available with compressed gas can be obtained, and that truly constant piston velocity is easier to achieve experimentally than a constant gas pressure.

Shear stresses and shear rates may be derived from an analysis of flow in the capillary. Let the length and diameter of the capillary be L and D, respectively, as shown in figure 3.7a. Also, let the pressure drop over the capillary be from P just before entry to zero at exit. Consider the forces acting on the cylindrical element of melt coaxial with the capillary shown in figure 3.7b. Its radius is r and length dz, z being the axial coordinate in the direction of flow. The pressures acting on the upstream and downstream faces of the element are p and $p + dp$, respectively, while the shear stress acting on its sides in the z direction is τ. These choices of pressure increment and shear-stress direction are made deliberately to be consistent with later more general analyses; here, both dp and τ are negative. Assuming that pressures and viscous shear forces are the only significant ones acting on the element, for steady flow they must be in equilibrium

$$(p + \mathrm{d}p)\pi r^2 = p\pi r^2 + 2\pi r\tau\, \mathrm{d}z$$

Hence

$$\frac{\mathrm{d}p}{\mathrm{d}z} \equiv P_z = \frac{2\tau}{r} \tag{3.9}$$

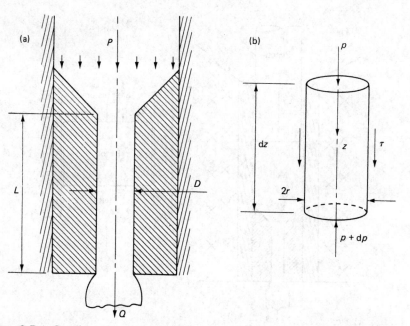

Figure 3.7   *Capillary rheometer analysis: (a) capillary geometry and operating conditions; (b) typical cylindrical element in the melt flow*

where P_z is a shorthand notation for the axial pressure gradient. The shear stress is therefore given by

$$\tau = \frac{1}{2} r P_z, \qquad P_z \approx -\frac{P}{L} \tag{3.10}$$

The P_z approximation makes no allowance for pressure losses at the entrance to the capillary. Note that, while the shear stress in the cone-and-plate rheometer was constant, in the capillary rheometer it is directly proportional to radius. This result is, however, true for any material, either newtonian or non-newtonian. For the purpose of deriving viscosity data it is convenient to take the shear stress at the wall of the capillary

$$\tau' = \frac{1}{4} P_z D \tag{3.11}$$

Now, since shear stress varies within the flow, then so must the shear rate. For a non-newtonian material, however, this variation is not linear and not known in advance. Therefore, it is usual to make the deliberate assumption that the melt is newtonian, and so derive the corresponding *apparent* shear rate and viscosity. These can then be corrected to find the true values.

The gradient of the axial velocity, w, with respect to the radius is

$$\frac{dw}{dr} = \frac{\tau}{\mu} = \frac{r P_z}{2\mu} \tag{3.12}$$

where μ is the assumed constant viscosity. Given the boundary condition $w = 0$ at $r = D/2$, this equation may be integrated to give the velocity profile

$$w = \frac{P_z}{4\mu} \left(r^2 - \frac{D^2}{4} \right) \tag{3.13}$$

Hence, the volumetric flow rate may be found as

$$Q = \int_0^{D/2} 2\pi r w \, dr = -\frac{\pi P_z D^4}{128\mu} \tag{3.14}$$

where the negative sign appears because P_z is negative. Using this result, the apparent shear rate at the wall of the capillary may be expressed in terms of the flow rate

$$\gamma_a' = \frac{\tau'}{\mu} = \frac{P_z D}{4\mu} = -\frac{32Q}{\pi D^3} \tag{3.15}$$

The apparent viscosity may be obtained as the ratio between τ' and γ_a'.

The lower limit of the shear rate range obtainable with a capillary rheometer is determined by the minimum pressure drop that can be accurately measured. On the other hand, the upper limit is determined by the maximum pressure that can be generated. At high pressure drops, however, significant errors tend to be introduced, both by the pressure dependence of viscosity and by the generation of heat, which is discussed below.

In addition to correcting apparent data for non-newtonian flow effects, there are a number of other corrections that can be applied to improve the accuracy of the results. Firstly, if the pressure, P, is measured at the piston rather than at the capillary entrance, it will include the pressure drop over the varying amount of melt in the barrel. This source of error can be eliminated by always recording the pressure when the piston is at the same small distance from the capillary.

A further source of error is the finite pressure drop between the position of pressure measurement and the true entry to the capillary. This end effect can be eliminated by using two or more capillaries of the same diameter but different lengths. Figure 3.8 shows a typical set of pressure readings obtained using three different capillary lengths, L_1, L_2 and L_3, and the same volumetric flow rate. Pressure is plotted against the ratio of capillary length to diameter. The three points should lie on a straight line. The main reason for any lack of linearity is that melt viscosity is significantly pressure dependent, and the mean pressure in the capillary increases with its length. Indeed, this approach has been used to estimate pressure dependence, although the results are not very accurate. If the experimental points are connected by a straight line as shown, which is then extrapolated to cross both axes, the pressure drop, P_0, associated with a zero-length capillary can be obtained, together with the ratio between the effective extra length of the capillary, e, due to end effects and the diameter. The true pressure gradient in the capillary may then be obtained in one of two ways

$$|P_z| = \frac{P - P_0}{L} = \frac{P}{L + e} \qquad (3.16)$$

where P and L are the values associated with any one of the capillaries. End corrections, e, can be as high as five to ten diameters. This fact justifies the assertion made in the last subsection that the capillary used in the melt flow index test, which has a length-to-diameter ratio of less than four, is too short to allow reliable values of the viscosity to be derived.

Generally, the most important correction that needs to be applied to capillary rheometer data is that due to non-newtonian flow effects. The usual method

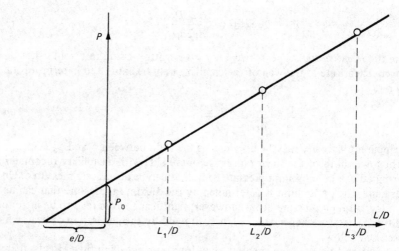

Figure 3.8   *Variation of pressure drop with capillary length-to-diameter ratio*

employed is known as the Rabinowitsch correction, and may be derived as follows. Integral equation 3.14 for the flow rate can be rearranged to give

$$Q = \int_0^{D/2} 2\pi r w \, dr = \int_0^{(D/2)^2} \pi w \, dr^2$$

Integration by parts yields

$$Q = [\pi w r^2]_0^{D/2} - \int_0^{D/2} \pi r^2 \frac{dw}{dr} \, dr$$

Provided the flow velocity is zero at the capillary wall, the first term in this equation is zero. Now, equations 3.10 and 3.11 give both the general expression for shear stress and the particular one for stress at the wall. Using these results, the radius can be eliminated from the above expression for flow rate to give

$$Q = - \int_0^{\tau'} \pi \left(\frac{D}{2} \frac{\tau}{\tau'} \right)^2 \frac{dw}{dr} \frac{D}{2\tau'} \, d\tau$$

which, after some manipulation, yields

$$\frac{8Q\tau'^3}{\pi D^3} = \frac{\gamma_a'\tau'^3}{4} = \int_0^{\tau'} \tau^2 \frac{dw}{dr} \, d\tau$$

Differentiating with respect to τ'

$$\frac{3}{4} \gamma_a' \tau'^2 + \frac{\tau'^3}{4} \frac{d\gamma_a'}{d\tau'} = \tau'^2 \gamma_t'$$

where γ_t' is the true shear rate at the capillary wall, which can now be expressed in terms of the apparent rate there as follows

$$\gamma_t' = \gamma_a' \frac{3n' + 1}{4n'} \quad , \qquad n' = \frac{d(\ln \tau')}{d(\ln \gamma_a')} \tag{3.17}$$

The purpose of this rather complicated analysis to avoid making any assumptions about the relationship between shear stress and shear rate. The only assumption made is the no-slip boundary condition at the capillary wall. If the plotted apparent data follow a power-law relationship of the form defined by equation 3.3, then $n' = n$ and the correction factor for converting apparent to true shear rates is a constant. The value of this factor for a typical power-law index of 0.5 is 1.25 and increases to 1.75 for $n = 0.25$. True viscosities can be obtained as the ratios between shear stresses and true shear rates.

Another potential source of error in capillary rheometer measurements is the heat generated in the flow. Although this effect cannot be eliminated, it is possible to estimate whether it is likely to be significant. The rate of dissipation of mechanical energy is PQ. Assuming adiabatic flow, this energy is retained in the melt and is therefore equal to the increase in thermal energy $\rho Q C_p \Delta T$, where ρ and C_p are the melt density and specific heat and ΔT is the mean temperature rise. Hence, this rise may be estimated as

$$\Delta T = \frac{P}{\rho C_p} \tag{3.18}$$

and depends only on the pressure drop and material properties. If the temperature rise so calculated is, say, less than 1 °C, the generated heat has a negligible effect. On the other hand, if it is of the order of 10 °C or more, the accuracy of the results is seriously affected.

The final major source of error in capillary rheometer data is that of slip between the melt and capillary wall, which invalidates the boundary condition used in deriving both equations 3.13 and 3.17. There is some evidence that most, if not all, polymer melts suffer slip at solid flow boundaries above some critical shear stress. Polyvinyl chloride is perhaps the most important example, if only because it appears to slip at relatively low shear stresses. Slip must be suspected in capillary-rheometer data if the power-law index appears to be very small — say, less than 0.1 — particularly if there is a relatively abrupt change from a higher value as the shear stress is increased. Flow curves of this type are sometimes reported for polyvinyl chloride. The very low value of n implies that the volumetric flow rate is almost independent of shear stress at the capillary wall: slip occurs and the melt flows as a plug. The viscosity data are meaningless.

3.3.4 Comparison of Methods of Viscosity Measurement

The practical advantages and disadvantages of cone-and-plate and capillary methods of measuring polymer melt viscosity have already been discussed. It is appropriate now to compare the two approaches in terms of their ability to establish viscosity data over the range of interest for processing operations. Figure 3.9 shows a typical

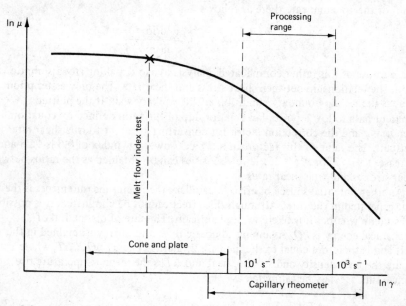

Figure 3.9   *Typical flow curve and rheometer shear rate ranges*

flow curve, with the shear rate range $10^1 - 10^3$ s^{-1} relevant to processing operations marked. Also indicated are the typical ranges of useful shear rate associated with cone-and-plate and capillary rheometers. While the latter includes the processing range, the maximum useful shear rate normally obtainable in cone-and-plate instruments often falls short of the processing range. There is usually a small overlap of the ranges obtainable with the two methods, which allows some checking of viscosity values. It should be noted that the melt flow index test is normally conducted at a shear rate that is well below the processing range.

Figure 3.9 demonstrates that, while rotational methods of measuring viscosity are useful for studying and comparing the flow properties of polymer melts, it is capillary methods that are normally employed to gather viscosity data for use in calculations concerned with processing operations. Melt flow indices should be used with caution, to compare the flow properties of similar polymers. Because they are obtained at low shear rates, they may give a misleading comparison of the properties under processing conditions of dissimilar materials. An important application of the melt flow index is in studying small variations in the viscosity of a nominally homogeneous material. For example, variations in the degree of degradation produced by a process can be monitored in terms of the corresponding increase in melt flow index.

3.4 Elastic Properties of Polymer Melts

So far, polymer melts have been treated as purely viscous fluids, whereas in practice they exhibit significant elastic properties. Such properties may be observed in both the cone-and-plate and capillary methods of measuring viscosity. Figure 3.10 shows the variations in stress and strain associated with an instrument of the type illustrated in figure 3.4. Torque is rapidly applied to the cone at time zero and abruptly removed at some later time when steady conditions have been achieved. The resulting shear-strain profile shows an initially rapid increase, settling down to the constant strain rate that is used for deriving viscosity. When the torque is removed, however, the direction of rotation of the cone is reversed since the material tends to recover its initial state. Thus, elasticity in melt flow exhibits itself as a fading memory effect. Shear strains ϵ_1 and ϵ_2 marked in figure 3.10 are usually called the initial elastic strain and the recovered elastic strain, respectively. The more elastic the material, the larger are these strains.

Elasticity in capillary flow shows itself as the phenomenon of die swell. Figure 3.11a shows a diagrammatic cross section of melt flowing from a rheometer barrel through a capillary of diameter D into the atmosphere, where it expands to a diameter of D'. The swelling ratio is defined as $S = D'/D$, and in practice normally lies in the range $1-2$. Also shown as dashed lines are some typical streamlines, which serve to emphasise the way in which material is forced to contract laterally in order to pass through the capillary before tending to recover elastically to its previous form.

In addition to elastic effects, there is a substantial change in velocity profile as the melt leaves the die. Typical forms of profile in the capillary and extrudate are indicated in the diagram. In the absence of elasticity and gravity, a viscous newtonian fluid would, as a result of the change in velocity profile, exhibit a

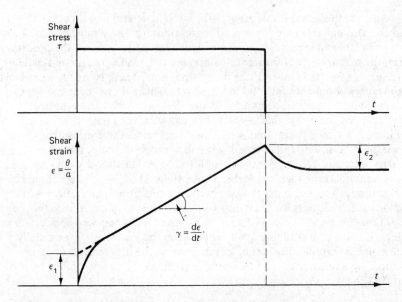

Figure 3.10   *Variations of shear stress and strain with time in a cone-and-plate rheometer*

swelling ratio of about 1.1. It must be emphasised, however, that this would only be the case if the flow is dominated by pressure and viscous forces as it is in the case of polymer melts. If inertia effects are important, then newtonian fluids can exhibit a contraction in diameter of an emerging jet, known as a *vena contracta*.

Figure 3.11*b* shows a typical plot of swelling ratio against flow rate for a polymer melt, using a capillary of a given length. At very low flow rates, the material behaves as a newtonian fluid with the appropriate swelling of about ten per cent. As Q is increased, however, the residence time in the capillary is decreased, allowing the melt to recover to a greater degree its former state. Above a certain rate of flow, however, the surface of the extrudate becomes distorted and it is no longer possible to measure its diameter. This phenomenon, which is most commonly known as *melt fracture*, may be at least partially elastic in origin. It also appears to be associated with intermittent slip between the melt and capillary wall of the type discussed at the end of section 3.3.3. Alternate local slipping and sticking at the capillary wall, aggravated by melt elasticity, provides a possible mechanism for the onset of extrudate roughness.

Figure 3.11*c* shows a typical plot of swelling ratio against capillary length for a given rate of flow. From a large degree of swelling associated with a short residence time for small capillaries, S decreases asymptotically to the newtonian value as the capillary length is increased. The die swell phenomenon is of considerable importance in processing operations involving flow from fixed to free boundary conditions. Swelling ratio data gathered from capillary rheometer experiments can form the basis of estimates of swelling under processing conditions. Melt fracture is also important, if only in the sense that it is normally to be avoided. As it only

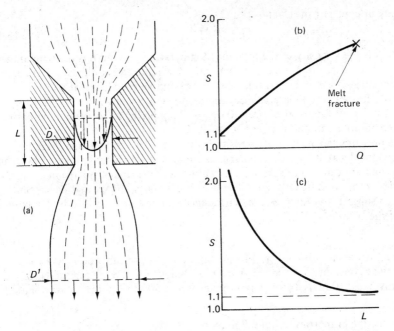

Figure 3.11   *Die swell in capillary flow: (a) extrudate dimensions, velocity profiles and flow stream lines; (b) variation of swelling ratio with flow rate; (c) variation of swelling ratio with capillary length*

occurs above a certain critical capillary wall shear stress, it is important to determine the magnitude of this stress experimentally.

3.5 Temperature and Pressure Dependent of Melt Properties

All polymer properties are at least somewhat dependent on temperature and pressure. Mainly as a result of experimental difficulties, it is usually only the forms of dependence of density and viscosity that are measured. Although the density of a polymer melt at processing temperature is significantly different from that of the same material at ambient temperature, it is not necessary to allow for local variations in density due to temperature variations within a melt flow. Similarly, although polymers are compressible, the change in volume is only significant under the levels of pressure involved in processes such as injection moulding.

Polymer melt viscosities are sensitive to changes in temperature. If flow curves of the type shown in figure 3.3 for a particular material are obtained at different temperatures, they are found to be nearly parallel to each other, displaced downwards by an amount proportional to the increase in temperature. Such an effect, which implies that the power-law index is not significantly dependent on temperature, can be represented by the following temperature dependence of viscosity at constant shear rate

$$\mu = \mu_0 \exp[-b(T - T_0)] \tag{3.19}$$

where μ_0 is the viscosity at the reference temperature T_0, and b is the temperature coefficient of viscosity at constant shear rate. Owing to the nonlinear relationship between shear stress and shear rate, the corresponding coefficient at constant shear stress is different (for a material obeying the power-law equation 3.3 it is b/n). Typically, b is of the order of $0.01-0.02 \ °C^{-1}$, implying a reduction in viscosity of between about 10 and 20 per cent for a 10 °C rise in temperature.

The pressure dependence of viscosity is more difficult to measure. Probably the most practical way is to use a capillary instrument in which the cavities at both ends of the capillary can be pressurised, giving a mean pressure that is large in comparison with the drop across the capillary. The limited amounts of data that have been obtained suggest that the pressure dependence can best be described by a relationship of the form

$$\mu = \mu_0 \exp(\alpha p) \tag{3.20}$$

where μ_0 is the viscosity at zero pressure and α is the pressure coefficient of viscosity at constant shear rate. Typical reported values of α are 5 m^2/GN for polyethylenes, 10 m^2/GN for polypropylenes and 50 m^2/GN for polystyrenes.

3.6 Processing Properties of Solid Polymers

Although the properties of polymers in the molten state are the most important for processing operations, some solid-state properties also influence such operations. Clearly, thermal properties such as conductivity and specific heat affect the ease with which solid polymers can be melted. The properties that most affect the flow of solid polymers in either granular or powder form are those of friction, both internal between the particles and external at the interfaces between the polymer and the boundaries confining the flow. Internal friction is important where the material flows freely, typically under the influence of gravity, with relative motion of the particles. One of the best examples in the polymer processing context is the flow in the hoppers of extruders and injection moulding machines. Once in a screw channel, the particles are compacted into a solid plug whose motion is then largely determined by the coefficients of friction at the screw and barrel surfaces.

The internal-frictional properties of particulate solid are commonly measured and compared in terms of the angle of repose. This is the angle formed by the free surface of the particles when heaped on a horizontal plane, and is directly related to the coefficient of internal friction. The magnitude of the friction coefficient at a rigid boundary is more difficult to determine. This is because it depends not only on the given combination of polymer and metal surface, but also on the surface temperature and the time for which relative motion occurs. The coefficient for a clean metal surface is generally much lower than the steady value obtained when the surface has become thoroughly coated with polymer. The time required for this transition increases with the hardness of the polymer and, in some cases, may be of the order of several hours. Clearly, both the temperature- and time-dependence effects are potentially very important in the solid feeding characteristics of processing equipment.

4

Fundamentals of Polymer Melt Flow

In the last chapter, the properties of polymers that influence their processing characteristics were considered. As processing operations normally involve deformation in the molten state, the main emphasis was on melt flow properties. It is now appropriate to establish the mathematical equations that can be used to describe the flow of polymer melts. In later chapters, these governing equations are applied, first to the analysis of relatively simple flows of the types occurring in, for example, extrusion dies, and then to more complex flows associated with screw extrusion and injection moulding.

For the purposes of most analyses of polymer processes, melts can be treated as continuous media and the equations of continuum mechanics employed. This is a reasonable approach provided the scale of interest, such as the depth of a flow channel or the thickness of a polymer film, is large compared with the typical molecule size. At such a scale of interest, physical properties can be regarded as continuous functions of position within the material. The definition of a property such as melt viscosity, already introduced in chapter 3, is only possible if the continuum assumption is adopted.

4.1 Tensor Notation

Before considering the fundamental equations of continuum mechanics, it is worth reviewing a system of tensor notation that is both conveniently compact and widely used for expressing such equations. For the benefit of readers not familiar with this notation, specific examples of general tensor equations will be displayed in full. Also, tensor notation will be discarded when particular problems are analysed in detail.

4.1.1 Suffix Notation

The principle of classifying physical quantities as being either scalars, vectors or tensors should be a familiar one. *Scalars* have magnitude only and are displayed without suffixes; for example, pressure, p, and temperature, T. *Vectors* have both magnitude and direction and are displayed with single suffixes; for example, velocity, v_i. *'Tensors'* have magnitude, direction and plane on which to act and are displayed with double suffixes; for example, viscous stress, τ_{ij}, and rate of deformation, e_{ij}. 'Tensors', as defined here, are really second-order tensors since higher orders having more suffixes exist. Similarly, vectors and scalars may be thought of as first- and zero-order tensors, respectively.

Of the suffixes of a tensor, such as stress τ_{ij}, the first refers to its direction while the second refers to the direction of the outward normal to the plane on which it acts. Thus, a particular component with $i = j$ is a direct stress and, if positive, is tensile, while a component with $i \neq j$ is a shear stress. The general directional suffix i refers to the coordinate direction x_i, which is not necessarily a cartesian coordinate, and may take the specific values 1, 2 or 3 corresponding to the coordinate directions x_1, x_2 or x_3. In cartesian systems, these directions, and the corresponding suffixes, are often displayed as x, y and z, while in, for example, cylindrical polar coordinates they become r, θ and z. While in general coordinates the velocity components are v_1, v_2 and v_3, for present purposes they become u, v and w in either cartesian or cylindrical polar systems.

4.1.2 Summation Conventions

The display of equations involving tensor quantities may be abbreviated with the aid of the following two rules.

(1) Any suffix appearing once only in a term of a tensor equation may take any value in its range.

(2) A suffix repeated in a term implies a summation of that term over the range of values of the suffix.

For example,

$$v_i = 0 \quad \text{means } v_1 = 0, v_2 = 0, v_3 = 0$$

$$e_{ii} = 0 \quad \text{means } e_{11} + e_{22} + e_{33} = 0$$

$$y_i = l_{ij}x_j \text{ means } y_1 = l_{11}x_1 + l_{12}x_2 + l_{13}x_3$$
$$y_2 = l_{21}x_1 + l_{22}x_2 + l_{23}x_3$$
$$y_3 = l_{31}x_1 + l_{32}x_2 + l_{33}x_3$$

4.1.3 Total and Viscous Stresses

While the *total* stress tensor, t_{ij}, is the one normally employed in the stress analysis of solid continua, in fluid-mechanics applications it is usual to separate the pressure, p, and define a *viscous* stress tensor, τ_{ij}

$$\tau_{ij} = t_{ij} + p\delta_{ij} \tag{4.1}$$

where δ_{ij} is the Kronecker delta

$$\delta_{ij} = 1 \text{ if } i = j, \; \delta_{ij} = 0 \text{ if } i \neq j$$

Tensor equation 4.1 summarises a total of nine relationships, such as

$$\tau_{11} = t_{11} + p, \tau_{12} = t_{12}$$

in a general coordinate system (x_1, x_2, x_3), or

$$\tau_{xx} = t_{xx} + p, \tau_{xy} = t_{xy}$$

in the cartesian system (x, y, z).

The pressure, p, is difficult to define rigorously, but for a stationary fluid it becomes the hydrostatic pressure. In practical terms, p may be taken as the 'pressure' measured by the usual methods, commonly at a flow boundary. It should be noted that, in order for rotational equilibrium to be maintained, both stress tensors must be symmetric

$$t_{ij} = t_{ji}, \tau_{ij} = \tau_{ji} \tag{4.2}$$

In other words, shear stresses are complementary.

4.1.4 Rate-of-deformation Tensor

The rate of deformation of a fluid at a particular point depends on the local velocity gradients, which may be expressed in tensor form as

$$g_{ij} = \frac{\partial v_i}{\partial x_j} \tag{4.3}$$

Displayed in full for cartesian and cylindrical polar coordinate systems

$$g_{ij} = \begin{bmatrix} \dfrac{\partial u}{\partial x} & \dfrac{\partial u}{\partial y} & \dfrac{\partial u}{\partial z} \\[2ex] \dfrac{\partial v}{\partial x} & \dfrac{\partial v}{\partial y} & \dfrac{\partial v}{\partial z} \\[2ex] \dfrac{\partial w}{\partial x} & \dfrac{\partial w}{\partial y} & \dfrac{\partial w}{\partial z} \end{bmatrix}, g_{ij} = \begin{bmatrix} \dfrac{\partial u}{\partial r} & \left(\dfrac{1}{r}\dfrac{\partial u}{\partial \theta} - \dfrac{v}{r}\right) & \dfrac{\partial u}{\partial z} \\[2ex] \dfrac{\partial v}{\partial r} & \left(\dfrac{1}{r}\dfrac{\partial v}{\partial \theta} + \dfrac{u}{r}\right) & \dfrac{\partial v}{\partial z} \\[2ex] \dfrac{\partial w}{\partial r} & \dfrac{1}{r}\dfrac{\partial w}{\partial \theta} & \dfrac{\partial w}{\partial z} \end{bmatrix} \tag{4.4}$$

Now, g_{ij} can be arranged as the sum of symmetric and antisymmetric parts

$$g_{ij} = \tfrac{1}{2}(g_{ij} + g_{ji}) + \tfrac{1}{2}(g_{ij} - g_{ji})$$
$$= e_{ij} + \omega_{ij} \tag{4.5}$$

where e_{ij} is the rate-of-deformation tensor, and ω_{ij}, which for present purposes is of much less importance, is the rate-of-rotation tensor.

4.2 Continuum Mechanics Equations

In any continuum mechanics analysis of a fluid flow problem, it is necessary to satisfy the conservation equations, namely the equations of conversation of mass, momentum and energy, and also the constitutive equation defining the behaviour of the material, and finally the boundary conditions for the particular problem. While the conservation and constitutive equations are expressed here in tensor notation, and some examples given for cartesian and cylindrical polar coordinates, the list of further reading should be consulted for sources displaying them in full for a number of different coordinate systems.

4.2.1 Conservation of Mass

The general differential equation of mass conservation, often known as the continuity equation, is

$$\frac{\partial \rho}{\partial t} + \frac{\partial}{\partial x_i}(\rho v_i) = 0 \tag{4.6}$$

where ρ is the melt density, and t is time. As indicated in section 3.5, it is not necessary to allow for local variations in density due to temperature and pressure variations. If ρ is constant, equation 4.6 becomes

$$g_{ii} = e_{ii} = 0 \tag{4.7}$$

The particular forms of this result for cartesian and cylindrical polar coordinates are

$$\frac{\partial u}{\partial x} + \frac{\partial v}{\partial y} + \frac{\partial w}{\partial z} = 0 \tag{4.8}$$

$$\frac{1}{r}\frac{\partial}{\partial r}(ru) + \frac{1}{r}\frac{\partial v}{\partial \theta} + \frac{\partial w}{\partial z} = 0 \tag{4.9}$$

4.2.2 Conservation of Momentum

The general differential equations of momentum conservation, or equilibrium, often known as the Navier–Stokes equations, are

$$\rho\frac{\partial v_i}{\partial t} + \rho v_j g_{ij} = \frac{\partial t_{ik}}{\partial x_k} + \rho b_i \tag{4.10}$$

where b_i is the body force vector due, for example, to gravity. The terms on the left-hand sides of these equations represent inertia effects, while the stress derivative terms on the right-hand sides are due to viscous and pressure forces. In polymer melt flows, it is normally reasonable to assume that inertia effects and body forces are negligible, some justification for this assertion being provided in section 4.5. Hence, introducing viscous stresses with the aid of equation 4.1, equations 4.10 become

$$\frac{\partial \tau_{ik}}{\partial x_k} = \frac{\partial p}{\partial x_i} \tag{4.11}$$

representing balances between viscous and pressure forces in the flow.

In cartesian coordinates, equations 4.11 become

$$\frac{\partial \tau_{xx}}{\partial x} + \frac{\partial \tau_{xy}}{\partial y} + \frac{\partial \tau_{xz}}{\partial z} = \frac{\partial p}{\partial x} \tag{4.12}$$

for stress equilibrium in the x direction, together with two similar equations for the y and z directions. Similarly, in cylindrical polars, the equilibrium equations for the r, θ and z directions are

$$\frac{1}{r}\frac{\partial}{\partial r}(r\tau_{rr}) + \frac{1}{r}\frac{\partial \tau_{r\theta}}{\partial \theta} - \frac{\tau_{\theta\theta}}{r} + \frac{\partial \tau_{rz}}{\partial z} = \frac{\partial p}{\partial r} \tag{4.13}$$

$$\frac{1}{r^2}\frac{\partial}{\partial r}(r^2\tau_{\theta r}) + \frac{1}{r}\frac{\partial \tau_{\theta\theta}}{\partial \theta} + \frac{\partial \tau_{\theta z}}{\partial z} = \frac{1}{r}\frac{\partial p}{\partial \theta} \tag{4.14}$$

$$\frac{1}{r}\frac{\partial}{\partial r}(r\tau_{zr}) + \frac{1}{r}\frac{\partial \tau_{z\theta}}{\partial \theta} + \frac{\partial \tau_{zz}}{\partial z} = \frac{\partial p}{\partial z} \tag{4.15}$$

4.2.3 Conservation of Energy

The general differential equation of energy conservation for a fluid of constant density is

$$\rho C_p \left(\frac{\partial T}{\partial t} + v_i \frac{\partial T}{\partial x_i}\right) = \frac{\partial}{\partial x_i}\left(k\frac{\partial T}{\partial x_i}\right) + t_{ij}e_{ij} \tag{4.16}$$

where T is temperature, C_p is specific heat and k is thermal conductivity, material properties that were discussed in section 3.1. The terms on the left-hand side of this equation represent thermal convection effects, while those on the right-hand side are due to thermal conduction and the dissipation of mechanical work into heat. Often the only permissible assumptions are that the flow is steady, thereby eliminating the time-derivative term, and the thermal conductivity is constant. Hence, in cartesian coordinates, equation 4.16 becomes

$$\rho C_p \left(u\frac{\partial T}{\partial x} + v\frac{\partial T}{\partial y} + w\frac{\partial T}{\partial z}\right) = k\left(\frac{\partial^2 T}{\partial x^2} + \frac{\partial^2 T}{\partial y^2} + \frac{\partial^2 T}{\partial z^2}\right)$$

$$+ \tau_{xx}\frac{\partial u}{\partial x} + \tau_{yy}\frac{\partial v}{\partial y} + \tau_{zz}\frac{\partial w}{\partial z} + \tau_{xy}\left(\frac{\partial u}{\partial y} + \frac{\partial v}{\partial x}\right)$$

$$+ \tau_{yz}\left(\frac{\partial v}{\partial z} + \frac{\partial w}{\partial y}\right) + \tau_{zx}\left(\frac{\partial w}{\partial x} + \frac{\partial u}{\partial z}\right) \tag{4.17}$$

4.3 Constitutive Equations

The viscous properties of polymer melts were discussed in section 3.2, and methods for measuring them under conditions of simple shear flow described in section 3.3. In many processing operations, however, the flow is much more complex and a more general relationship between the states of stress and deformation is required. Such a relationship is known as the constitutive equation for the material. In view of the fact, discussed in section 3.4, that polymer melts exhibit elastic properties, the constitutive equation should be a viscoelastic one, relating stresses not only to current local rates of strain but also to the strain history of the particular element of material. Although many viscoelastic constitutive equations have been proposed, they are generally difficult to use, not least because of the practical problems of measuring the material properties required. In many polymer processes, however, elastic memory effects are not very important, because the melts are subjected to large steady rates of deformation for relatively long times. This is particularly true of channel flows of the types encountered in, for example, extruder screws and dies, and is given further consideration in section 4.6.

4.3.1 The Stokesian Fluid

The most general constitutive equation for an inelastic, homogeneous and isotropic fluid is that of the stokesian fluid

$$\tau_{ij} = \eta_1 e_{ij} + \eta_2 e_{ik} e_{kj} \tag{4.18}$$

where η_1 and η_2 are the generalised viscosity and cross viscosity, respectively. Note that τ_{ij} depends only on rates of deformation obtained from the symmetric part of the velocity gradient tensor in equation 4.5, and not on rates of rotation. In cartesian coordinates, for example, equation 4.18 yields for a typical stress component

$$\tau_{xy} = \eta_1 e_{xy} + \eta_2 (e_{xx} e_{xy} + e_{xy} e_{yy} + e_{xz} e_{zy}) \tag{4.19}$$

The material properties η_1 and η_2 of a stokesian fluid depend on the local state of deformation rate and also on the thermodynamic variables, temperature and pressure. Clearly, the state of deformation rate should be independent of the particular coordinate system used to describe it. There are three, and only three, scalar parameters that define any state of deformation rate independently of the coordinate system. These are the principal invariants of the rate-of-deformation tensor

$$I_1 = e_{ii} \tag{4.20}$$

$$I_2 = \tfrac{1}{2} e_{ij} e_{ij} \tag{4.21}$$

$$I_3 = \det (e_{ij}) \tag{4.22}$$

where 'det' means the determinant of the enclosed matrix. In cartesian coordinates, for example

$$I_1 = e_{xx} + e_{yy} + e_{zz} \tag{4.23}$$

$$I_2 = \tfrac{1}{2}(e_{xx}^2 + e_{yy}^2 + e_{zz}^2) + e_{xy}^2 + e_{yz}^2 + e_{zx}^2 \tag{4.24}$$

$$I_3 = \begin{vmatrix} e_{xx} & e_{xy} & e_{xz} \\ e_{yx} & e_{yy} & e_{yz} \\ e_{zx} & e_{zy} & e_{zz} \end{vmatrix} \tag{4.25}$$

Some physical significance can be attached to these invariants. Clearly I_1 represents change in volume, which is zero for a material of constant density according to equation 4.7. The second invariant, I_2, represents a mean rate of deformation including all tensile and shear components, and is the most important invariant for constitutive equation purposes. Finally, I_3 is zero in the absence of tensile deformation rates, and can therefore be thought of as a measure of the importance of tensile deformation in a particular flow.

A simple illustration of the invariance of I_1, I_2 and I_3 is provided by the shear flow between flat parallel boundaries shown in figure 4.1. Figure 4.1a shows the flow due to movement of the upper boundary in some arbitrary direction, s, to the x and z axes, which are in the plane of the lower boundary, coordinate y being normal to both surfaces. The components of the upper boundary velocity are V_x and V_z in the x and z directions, respectively, with resultant V_s. Given that the boundaries are a uniform distance H apart, the velocity components within the flow are

$$u(y) = y\,\frac{V_x}{H}, \; v = 0, \; w(y) = y\,\frac{V_z}{H} \tag{4.26}$$

Hence, using equations 4.23–4.25, $I_1 = 0$ and

$$I_2 = \frac{1}{4}\left[\left(\frac{V_x}{H}\right)^2 + \left(\frac{V_z}{H}\right)^2\right] = \frac{1}{4}\left(\frac{V_s}{H}\right)^2 \tag{4.27}$$

$$I_3 = \begin{vmatrix} 0 & \tfrac{1}{2}V_x/H & 0 \\ \tfrac{1}{2}V_x/H & 0 & \tfrac{1}{2}V_z/H \\ 0 & \tfrac{1}{2}V_z/H & 0 \end{vmatrix} = 0 \tag{4.28}$$

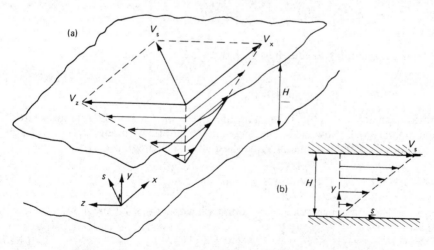

Figure 4.1 *Shear flow between flat surfaces used to illustrate invariance: (a) general view; (b) view of the plane of resultant motion*

In contrast, figure 4.1*b* shows the same resultant flow in the y, s plane, with a linear velocity profile increasing from zero to V_s at the upper boundary. Hence, using coordinate s in preference to x and z, I_1 and I_3 are again zero and

$$I_2 = e_{sy}{}^2 = \frac{1}{4}\left(\frac{V_s}{H}\right)^2 \qquad (4.29)$$

which is identical to equation 4.27. In other words, the choice of coordinates does not affect the values of the three scalar invariants.

Returning to the generalised viscosity and cross viscosity, these can now be expressed in functional-dependence form as

$$\eta_{1,2} = \eta_{1,2}(I_1, I_2, I_3, T, p) \qquad (4.30)$$

For a constant-density material, $I_1 = 0$ and is therefore not a variable, and

$$\eta_{1,2} = \eta_{1,2}(I_2, I_3, T, p) \qquad (4.31)$$

It should be noted that the familiar newtonian-fluid constitutive equation is merely a special case of the stokesian fluid in which $\eta_2 = 0$ and η_1 is independent of the state of deformation rate.

4.3.2 Correlation with Viscosity Measurements in Simple Shear Flow

Having selected a general constitutive equation for polymer melts, the material parameters η_1 and η_2 must be correlated as far as possible with the viscosity data obtained from capillary rheometer measurements as described in section 3.3.3. If the cylindrical polar coordinates shown in figure 3.7 are used to describe the uniform capillary flow then, according to equations 4.4 and 4.5, the only nonzero velocity gradient and rates of deformation are

$$g_{zr} = \frac{dw}{dr}, \, e_{rz} = e_{zr} = \tfrac{1}{2}\frac{dw}{dr} \qquad (4.32)$$

Therefore, according to equations 4.21 and 4.22, $I_3 = 0$ and

$$I_2 = \frac{1}{4}\left(\frac{dw}{dr}\right)^2 = \frac{1}{4}\gamma^2 \qquad (4.33)$$

at the capillary wall; γ ($\gamma_t{}'$ in the notation of section 3.3.3) is the true shear rate at the capillary wall. Now, with only one nonzero deformation rate, $e_{zk}e_{kr} = 0$, and equations 4.18 and 4.31 defining the shear stress at the wall become

$$\tau \equiv \tau_{zr} = \eta_1\left(\frac{1}{4}\gamma^2, 0, T, p\right)\frac{\gamma}{2} \qquad (4.34)$$

irrespective of the magnitude of η_2. Comparing equations 3.2 and 4.34, it is clear that

$$\eta_1 \equiv 2\mu \qquad (4.35)$$

In other words, the generalised viscosity normally used in the stokesian constitutive equation is simply twice the more familiar shear viscosity.

According to equation 3.3, there is very often an empirical power-law relationship between shear stress and shear rate measured in simple shear flows, of the form

$$\tau = C\gamma^n \tag{4.36}$$

As previously discussed in section 3.2, this is not a very convenient form for mathematical manipulation because the units of C depend on n, and negative shear rates are not admissible. Comparing equations 4.34 and 4.36, however

$$\mu = \frac{\tau}{\gamma} = C\gamma^{n-1} \equiv \frac{1}{2}\eta_1 \left(\frac{1}{4}\gamma^2, 0, T, p \right) \tag{4.37}$$

where T and p are the temperature and mean pressure of the test. A convenient form of generalisation of viscosity measurements obtained from simple shear flows is therefore

$$\tfrac{1}{2}\eta_1 = \mu = \mu_0 \left| \frac{\sqrt{(4I_2)}}{\gamma_0} \right|^{n-1} \tag{4.38}$$

Note the introduction of an effective viscosity, μ_0, at some reference shear rate, γ_0, to overcome the objection to the variable dimensions of C. Also, the replacement of γ by the positive second invariant avoids the difficulty of negative shear rate. Finally, the temperature and pressure dependence of viscosity can be included with the aid of empirical equations 3.19 and 3.20.

4.3.3 Normal Stresses

When a polymer melt is subjected to a simple shear flow — for example, of the type shown in figure 4.1 — a direct stress component normal to the flow boundaries is generated. Although this is at least partly due to the elastic nature of the material, it is worth noting that even the purely viscous stokesian constitutive equation exhibits such an effect. For the flow shown in Figure 4.1b, the normal stress derived from equations 4.18 and 4.29 is

$$\tau_{yy} = \eta_2 e_{ys} e_{sy} = \frac{1}{4}\eta_2 \left(\frac{V_s}{H} \right)^2 \tag{4.39}$$

Thus, it is the cross-viscosity term of the constitutive equation, which is not present in simple newtonian fluids, which gives rise to normal stresses. In many practical processing operations, however, normal stresses are small compared with the pressures associated with melt flow along narrow channels.

4.3.4 Extensional Flow

Simple shear flows are of a particular type in which the third invariant of the rate-of-deformation tensor, I_3, is zero. While such flows predominate in many processes, particularly within enclosed channels, extensional flows involving tensile deformations are also important. The best examples are provided by free surface flows of the type encountered in the film blowing and fibre spinning processes.

Consider a uniform filament of melt lying in the x cartesian coordinate direction. Let the rate of extension of this filament be

$$e_{xx} = \frac{\partial u}{\partial x} = e \tag{4.40}$$

Hence, assuming the density of the melt to be constant, equation 4.7 can be used to obtain

$$e_{yy} = e_{zz} = -\tfrac{1}{2}e \tag{4.41}$$

all other deformation rates being zero. Using equations 4.24 and 4.25

$$I_2 = \frac{3}{4} e^2, I_3 = \frac{1}{4} e^3 \tag{4.42}$$

and, from equations 4.1 and 4.18, the total direct stresses are

$$t_{xx} = -p + \tau_{xx} = -p + \eta_1 e + \eta_2 e^2$$

$$t_{yy} = t_{zz} = -p - \eta_1 \frac{e}{2} + \eta_2 \frac{e^2}{4} = 0$$

Eliminating p

$$t_{xx} = \frac{3}{2} \eta_1 e + \frac{3}{4} \eta_2 e^2 \tag{4.43}$$

The Trouton extensional viscosity is defined as

$$\lambda = \frac{t_{xx}}{e_{xx}} = \frac{3}{2} \eta_1 + \frac{3}{4} \eta_2 e \tag{4.44}$$

where

$$\eta_{1,2} = \eta_{1,2} \left(\frac{3}{4} e^2, \frac{1}{4} e^3, T, p \right) \tag{4.45}$$

For a newtonian fluid, $\eta_2 = 0$ and

$$\lambda = \frac{3}{2} \eta_1 = 3\mu \tag{4.46}$$

showing that the viscosity in extensional flow is three times that in shear flow. Experiments on molten polymers have shown that, while the ratio λ/μ is approximately 3 at low rates of deformation, where the shear behaviour is nearly newtonian, at higher rates it may increase to 300 or more. This is because, although μ decreases with increasing rate of deformation, λ generally remains nearly constant and may even increase slightly. While this effect can be interpreted in terms of equation 4.44, it is not possible on the basis of shear and extensional flow measurements alone to distinguish the effect of the η_2 term from the dependence of η_1 on I_3. Clearly, the generalised power-law equation 4.38 derived from measurements in simple shear flow is only reliable for flows in which cross viscosity is unimportant and I_3 is small.

4.4 Boundary Conditions

Having considered the general differential equations governing polymer melt flows, it remains to specify the relevant boundary conditions before the solution for a particular problem can be obtained. Indeed, it is the boundary conditions as much as the equations themselves that determine the nature of this solution. For internal flows bounded by solid surfaces, the conditions associated with the equilibrium equations are usually expressed in terms of velocities. Assuming no slip at a solid–melt interface, the velocity of the melt there is equal, in magnitude and direction, to that of the boundary surface. As indicated in sections 3.3.3 and 3.4, however, there is some evidence that most, if not all, polymer melts suffer slip at solid flow boundaries above some critical shear stress. This should be borne in mind when attempting to specify boundary conditions for regions of intense shear – for example, in or near the clearance between an extruder screw flight and barrel. For free surface flows, and internal flows when slip is occurring, the boundary conditions are expressed in terms of stresses – for example, zero shear and normal stresses at a free surface.

The thermal boundary conditions associated with the energy conservation equation can be of different types. The simpler forms are those involving either fixed temperature of a boundary, or a known gradient normal to it due to a prescribed heat transfer rate. Making the reasonable assumption that there is good thermal contact at a solid–melt interface, the temperatures of the two media must be equal there. If n is the direction locally normal to a melt flow boundary, the most general form of thermal boundary condition is

$$a_1 \frac{\partial T}{\partial n} + a_2 T + a_3 = 0 \qquad (4.47)$$

where a_1, a_2 and a_3 are usually constants, but could be functions of temperature. For example, suppose h is the heat transfer coefficient between a free surface polymer boundary and the surrounding environment, the temperature of which is T_∞ remote from the surface. The heat transfer rate to the surroundings is

$$q = -k \frac{\partial T}{\partial n} = h(T - T_\infty) \qquad (4.48)$$

both T and its outward normal derivative in the melt being evaluated at the boundary. Clearly, this is of the general form displayed in equation 4.47. Similarly, the prescribed boundary temperature and temperature gradient conditions are merely special cases, obtained when a_1 and a_2 are zero, respectively.

4.5 Dimensional Analysis of Melt Flows

Dimensional analysis can be used to generalise solutions to the melt flow equations by combining the physical variables to form a smaller number of dimensionless parameters. The flow behaviour may then be characterised in terms of these parameters. This is particularly useful when comparing flows occurring under different physical conditions. A good example is provided by the problem of scaling

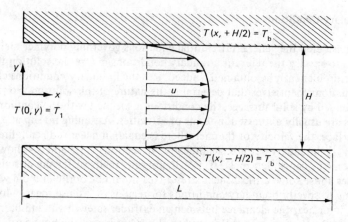

Figure 4.2 *A simple channel flow between flat parallel surfaces*

up the design of a piece of processing machinery from, say, a laboratory trial to larger production equipment.

Consider the simple channel flow between flat parallel surfaces shown in figure 4.2. This type of flow is found in many polymer processing operations — for example, in extrusion dies, screw channels and injection moulds. The boundaries are of length L in the direction of flow and a distance H apart, where usually $H \ll L$, and can be regarded as being of infinite extent in the z direction normal to the plane shown. Melt is admitted to the channel at, say, a uniform temperature T_1, and the flow boundaries are maintained at temperature T_b. Let coordinates x and y be chosen as shown, $y = 0$ being the mid-plane of the channel. Assuming that the flow is steady and occurs only in the x, y plane, as a result either of a pressure difference or of a relative velocity between the boundaries in the x direction, the functional forms of the velocity components and temperature are

$$u = u(x, y), \qquad v = v(x, y), \qquad w = 0, \qquad T = T(x, y) \quad (4.49)$$

Neglecting body forces, but retaining for the moment melt inertia effects, the momentum-conservation equation for the x direction may be derived from equation 4.10 as

$$\rho \left(u \frac{\partial u}{\partial x} + v \frac{\partial v}{\partial y} \right) = -\frac{\partial p}{\partial x} + \frac{\partial \tau_{xx}}{\partial x} + \frac{\partial \tau_{xy}}{\partial y} \quad (4.50)$$

While a similar equation would be required for the y direction to define the motion fully, for the purposes of dimensional analysis attention can be confined to equilibrium in the x direction.

The shear stresses may be obtained from equations 4.18 and 4.35 as

$$\tau_{xx} = 2\mu \frac{\partial u}{\partial x}, \qquad \tau_{xy} = \mu \left(\frac{\partial u}{\partial y} + \frac{\partial v}{\partial x} \right) \quad (4.51)$$

μ being defined by equation 4.38, together with an exponential temperature dependence of the form displayed in equation 3.19. This use of the simplified stokesian constitutive equation, which neglects cross-viscosity and third invariant effects, is justified by the fact that the flow is predominantly simple shear. Energy conservation equation 4.17 also becomes

$$\rho C_p \left(u \frac{\partial T}{\partial x} + v \frac{\partial T}{\partial y} \right) = k \left(\frac{\partial^2 T}{\partial x^2} + \frac{\partial^2 T}{\partial y^2} \right) + 4\mu I_2 \qquad (4.52)$$

where

$$I_2 = \frac{1}{2} \left(\frac{\partial u}{\partial x} \right)^2 + \frac{1}{2} \left(\frac{\partial v}{\partial y} \right)^2 + \frac{1}{4} \left(\frac{\partial u}{\partial y} + \frac{\partial v}{\partial x} \right)^2 \qquad (4.53)$$

Taking H as the characteristic length scale for the flow, dimensionless coordinates may be defined as

$$X = \frac{x}{H}, \qquad Y = \frac{y}{H} \qquad (4.54)$$

Similarly, dimensionless velocity components are

$$U = \frac{u}{\bar{U}}, \qquad V = \frac{v}{\bar{U}} \qquad (4.55)$$

where $\bar{U}$ is a characteristic flow velocity, in the x direction. For the flow illustrated in figure 4.2, in which the channel boundaries are stationary, $\bar{U}$ may be defined as the mean velocity, Q/H; Q is the volumetric flow rate per unit width of channel normal to the plane of the flow. If there is relative motion between the boundaries in the x direction, $\bar{U}$ may be more conveniently defined as the relative velocity. There are various ways in which a dimensionless temperature might be defined but, in view of the importance of heat generated by viscous dissipation in melt flows, and its effect on viscosities, the following form is one of the most convenient

$$T^* = b(T - T_1) \qquad (4.56)$$

where b is the temperature coefficient of viscosity introduced in equation 3.19.

In terms of the dimensionless variables, equations 4.50 and 4.52 become

$$Re \left(U \frac{\partial U}{\partial X} + V \frac{\partial V}{\partial Y} \right) = -\pi_P + \frac{\partial \tau_{xx}{}^*}{\partial X} + \frac{\partial \tau_{xy}{}^*}{\partial Y} \qquad (4.57)$$

$$Pe \left(U \frac{\partial T^*}{\partial X} + V \frac{\partial T^*}{\partial Y} \right) = \frac{\partial^2 T^*}{\partial X^2} + \frac{\partial^2 T^*}{\partial Y^2} + 4G \frac{\mu}{\bar{\mu}} I_2{}^* \qquad (4.58)$$

where

$$I_2{}^* = \left(\frac{H}{\bar{U}} \right)^2 I_2$$

$$\tau_{xx}{}^* = \frac{\tau_{xx}}{\bar{\tau}} = 2 \frac{\mu}{\bar{\mu}} \frac{\partial U}{\partial X}$$

$$\tau_{xy}{}^* = \frac{\tau_{xy}}{\bar{\tau}} = \frac{\mu}{\bar{\mu}} \left(\frac{\partial U}{\partial Y} + \frac{\partial V}{\partial X} \right)$$

and $\bar{\tau}$ and $\bar{\mu}$ are the mean shear stress and viscosity, respectively, evaluated at the mean shear rate, $\bar{\gamma} = \bar{U}/H$, and inlet temperature T_1. Hence

$$\frac{\mu}{\bar{\mu}} = (I_2{}^*)^{(n-1)/2} \exp(-T^*) \tag{4.59}$$

The parameters Re, π_P, Pe and G are, respectively, the Reynolds number, dimensionless pressure gradient, Peclet number and Griffith number for the flow

$$Re = \frac{\rho \bar{U} H}{\bar{\mu}} \tag{4.60}$$

$$\pi_P = \frac{H}{\bar{\tau}} \frac{\partial p}{\partial x} \tag{4.61}$$

$$Pe = \frac{\rho C_p \bar{U} H}{k} \tag{4.62}$$

$$G = \frac{b \bar{\tau} \bar{\gamma} H^2}{k} \tag{4.63}$$

Clearly, the Peclet number provides a measure of the importance of thermal convection, which in the present flow occurs predominantly in the x direction. There is therefore a good case for using L, the natural length of the channel, as the length scale for defining a new dimensionless coordinate, $A = x/L$, in place of X. Equation 4.58 then becomes

$$Gz \left[U \frac{\partial T^*}{\partial A} + V \frac{\partial T^*}{\partial Y} \left(\frac{L}{H}\right) \right] = \left(\frac{H}{L}\right)^2 \frac{\partial^2 T^*}{\partial A^2} + \frac{\partial^2 T^*}{\partial Y^2} + 4G \frac{\mu}{\bar{\mu}} I_2{}^* \tag{4.64}$$

where Gz is the Graetz number

$$Gz = \frac{\rho C_p \bar{U} H^2}{kL} \tag{4.65}$$

The range of integration of equation 4.64 is from $A = 0$ to $A = 1$, rather than the less natural $X = 0$ to $X = L/H$ associated with equation 4.58.

In order to solve equations 4.57 and 4.58 or 4.64, the initial and boundary conditions for U, V and T^* are required. The initial conditions take the forms

$$U = U_1(Y), \qquad V = V_1(Y), \qquad T^* = 0 \text{ at } X = A = 0 \tag{4.66}$$

and the boundary conditions for the case shown, where the boundaries are stationary, are

$$U = 0, \qquad V = 0, \qquad T^* = \frac{G}{Br} \text{ at } Y = \pm\tfrac{1}{2} \tag{4.67}$$

where Br is a Brinkman number for the flow

$$Br = \frac{\bar{\tau} \bar{\gamma} H^2}{k(T_b - T_1)} \tag{4.68}$$

Having formulated the governing equations and boundary conditions in terms of dimensionless variables and parameters, it is now possible to examine the nature of the solutions to particular physical problems in terms of these parameters. Consider first the equilibrium equation 4.57, in which the inertia and pressure gradient terms involve the multipliers Re and π_P, respectively, while the dimensionless viscous stress terms are not so scaled. Therefore, Re and π_P are measures of the importance of inertia and pressure forces relative to viscous forces. In all practical melt flows, Reynolds numbers are very small, typically of the order of 10^{-3}, and almost always very much less than unity. Consequently, melt flows are often described as slow, and are always laminar. Also, the assumption of negligible inertia effects introduced in section 4.2.2 is justified: the left-hand side of equation 4.57 may be set to zero.

Turning now to the dimensionless energy equations 4.58 and 4.64, the thermal-convection and viscous-dissipation terms involve the multipliers Pe or Gz and G, respectively, while the thermal-conduction terms remain unscaled. Therefore, Pe and Gz are measures of thermal convection, and G is a measure of heat generation, all relative in importance to thermal conduction. The additional thermal parameter Br enters the temperature analysis via the boundary conditions and determines the relative importance of viscous dissipation and imposed differences in boundary temperature in bringing about temperature changes in the flow.

Comparing equations 4.57 and 4.58, it is clear that Pe in the energy equation is analogous to Re in the equilibrium equation. While Re is very small, however, Pe is almost invariably very large, typically of the order of 10^3-10^5, implying that thermal convection is an important mode of heat transfer in melt flows. By virtue of the use of H as a length scale throughout equation 4.58, it is difficult to compare, for example, the relative importance of conduction in the X and Y directions. This difficulty is overcome by equation 4.64, from which it is clear that, for the typical situation where $H \ll L$, thermal conduction in the dominant direction of flow is negligible compared with conduction in the Y direction normal to the flow boundaries. Equation 4.64 also shows that Gz is a measure of the importance of thermal convection in the direction of flow relative to conduction normal to the flow. Although Gz is often large, in some melt flows it may be of order unity or less.

In addition to quantifying the ratio between heat generation and thermal conduction, the Griffith number also determines the extent to which temperature changes generated within the melt lead to changes in viscosity, and hence in the velocity profiles. It is clear from equation 4.58 and the boundary conditions given by equations 4.66 and 4.67 that if $G = 0$, $T^* = 0$ everywhere, and there is no coupling between the energy and equilibrium equations via equation 4.59. In other words, the flow is isothermal. The fact that $T^* = 0$ in isothermal flow is a consequence of the method chosen to nondimensionalise temperatures, and does not imply that there are no temperature changes, merely that they have no effect on viscosities. In practice, G may be relatively large, sometimes as high as 10^2 or more.

A number of different flow regimes can be distinguished for the purposes of analysis in terms of their heat-transfer parameters, in the following way.

(1) If both G and G/Br are substantially less than unity, and preferably of the order of 10^{-1} or less, the flow may be treated as isothermal and the velocity profiles determined without reference to the temperature changes either generated by the flow or imposed by the thermal boundary conditions.

(2) Irrespective of the value of G, if Gz is substantially less than unity, heat transfer in the flow is dominated by thermal conduction to and from the boundaries. In other words, the flow is thermally fully developed and the temperature profile does not change in the direction of flow.

(3) If Gz is of order unity, then both thermal convection and conduction are important. Not only should temperature profiles be treated as developing but, assuming $G > 1$, so should the velocity profiles, by virtue of the coupling between temperatures and velocities.

(4) If Gz is large, then thermal convection is the dominant mode of heat transfer, although conduction cannot be neglected. Temperature profiles develop very slowly in the direction of flow. Although this causes slight changes in velocity profiles through coupling, it is reasonable to assume locally fully developed velocities.

Of these regimes, the most commonly encountered in practice are (1) and (4), although the thermally fully developed condition associated with (2) may be approached towards the end of a long flow channel. Assuming that a Graetz number of the order of 10 is necessary for this to be achieved, irrespective of the conditions at the flow inlet, the required channel length may be estimated with the aid of equation 4.65 as

$$L' = \frac{\rho C_p \overline{U} H^2}{10k} \tag{4.69}$$

4.6 The Lubrication Approximation

Before seeking solutions to the melt-flow equations, they can be considerably simplified by introducing the lubrication approximation. Some justifications for this procedure have already been established by the above dimensional analysis of melt flow, although variations in channel geometry can also be accommodated. In essence, the lubrication approximation involves the local replacement of flow in a narrow gap between smooth — but not necessarily flat or parallel — surfaces by uniform fully developed flow between plane parallel surfaces. In this context, 'narrow' means that the gap is small compared with the lengths along the boundaries over which the gap varies significantly.

Let H and L again be the length scales in the gap and along the boundaries, as shown in figure 4.2, although H is now not necessarily constant. In terms of geometric effects, the lubrication approximation is valid provided

$$\left| \frac{\partial H}{\partial x} \right| \ll 1 \tag{4.70}$$

Theoretical and experimental investigations have shown that the limiting channel taper is of the order of $10°$ ($\partial H/\partial x = 0.2$). Assuming that this geometric condition is is satisfied, other effects that could invalidate the lubrication approximation include those due to melt inertia, elasticity and thermal convection. The relative importance of melt elasticity effects can be expressed in terms of a dimensionless memory time

$$t_m{}^* = t_m \frac{\overline{U}}{L} \tag{4.71}$$

where t_m is the local elastic memory time, and $\overline{U}$ is the characteristic melt velocity. As indicated in section 4.3, melts are usually subjected to large steady rates of deformation for relatively long times, and t_m* is small enough for elastic effects to be unimportant.

The validity of the lubrication approximation must be considered separately for velocity and temperature profiles. Because the Reynolds number as defined in the last section is very small, inertia effects are negligible. Therefore, the lubrication approximation may be applied to velocities unless they are strongly coupled to rapidly developing temperatures, which is rare in practice. The approximation may only be applied to temperatures if either the Graetz number is small, or the region of interest is remote from the inlet to the flow channel.

The simplifying effects of introducing the lubrication approximation can be illustrated with reference to the flow shown in figure 4.2. If velocities are treated as being locally fully developed, then equations 4.49 reduce to

$$u = u(y), v = 0, w = 0 \tag{4.72}$$

and equilibrium equation 4.50 becomes

$$\frac{d\tau_{xy}}{dy} = \frac{dp}{dx} \equiv P_x \tag{4.73}$$

where the pressure gradient P_x is the independent of y. Also, energy equation 4.52 reduces to

$$\rho C_p u \frac{\partial T}{\partial x} = k \frac{\partial^2 T}{\partial y^2} + 4\mu I_2 \tag{4.74}$$

$$I_2 = \frac{1}{4} \left(\frac{du}{dy}\right)^2 \tag{4.75}$$

If the temperatures may also be treated as locally fully developed, equation 4.74 becomes

$$0 = k \frac{d^2 T}{dy^2} + 4\mu I_2 \tag{4.76}$$

4.7 Mixing in Melt Flows

An important function of many melt-flow processes is to produce a homogeneous product. This is particularly true of screw extrusion for example. Since perfect uniformity is impossible to achieve, the degree of homogeneity is a major factor affecting product quality. Homogeneity is achieved by mixing in the flow and is particularly important when colourants, fillers or other additives are to be incorporated.

A precise quantitative definition of mixing and mixedness is difficult, and must in general be statistical owing to the random nature of mixing processes. Two quantities are, however, useful for describing mixedness, namely the scale of segregation and the intensity of segregation. The scale of segregation is a measure of the mean size of regions of the same component in the mixture, while the intensity of segregation is a measure of the difference in concentration of the relevant property, such as colour, between the components. This difference in concentration is affected by diffusion at a molecular level. Since such diffusion is negligible in melt flow processes, the intensity of segregation is constant. A melt can be said to be well mixed when the scale of segregation is less than the scale of examination, for example, the minimum size that can be resolved by eye.

The scale of segregation is affected by both dispersive and distributive mixing. Dispersive mixing involves rupturing the lumps of the various components of the mixture, and therefore depends primarily on the stresses in the material. On the other hand, distributive mixing depends on the total deformation, which spreads the lumps into thin layers, and is generally the more important mixing mechanism in polymer processes.

The degree of distributive mixing imparted by a process can be defined as the ratio between the initial and final thicknesses of striations of a component of the mixture

$$M = \frac{S}{S'} \tag{4.77}$$

Consider first the mixing imparted by a simple shear flow. Figure 4.3a shows an element of one fluid component of a mixture contained between two strips of a second minor component. These strips are of length h and lie parallel to the y axis, their distance apart in the x direction being S, which is therefore the initial thickness of the major-component striation. Now, suppose the mixture is subjected to the simple shear flow

$$\frac{du}{dy} = \gamma, \, u = 0 \text{ at } y = 0; \, v = 0 \tag{4.78}$$

and consider the deformed shape of the element after a time interval, as shown in figure 4.3b. While the striation thickness in the x direction is unchanged, the sides of the element are now inclined at an angle α to this direction. Therefore, the

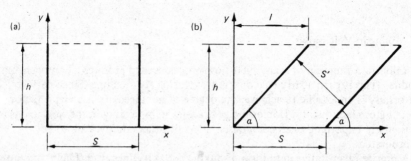

Figure 4.3 *The mechanism of distributive mixing in a simple shear flow;*
(a) initial orientation of a fluid element bounded by two strips of the minor
component of the mixture; (b) the deformed shape of the same element

striation thickness is

$$S' = S \sin \alpha \qquad (4.79)$$

Now, if the total shear strain is large

$$\sin \alpha \approx \tan \alpha = \frac{h}{l}$$

$$M = \frac{S}{S'} = \frac{l}{h} \qquad (4.80)$$

where l is the fluid displacement at $y = h$. But l/h is equal to the total shear strain imposed at rate γ. Hence, the mean rate of mixing is

$$\frac{dM}{dt} = \gamma \qquad (4.81)$$

for striations initially normal to the direction of shear. Although the mixing rate depends on the initial orientation, it is always proportional to γ. The shear rate may therefore be used to characterise the rate of mixing imparted to arbitrarily oriented components in a simple shear flow.

As the rate of mixing should not depend on the coordinate system chosen to describe a particular flow, it is to be expected that the generalised form of equation 4.81 for more complex flows will involve only the invariants of the rate-of-deformation tensor, particularly the second invariant. Hence

$$\frac{dM}{dt} = (4I_2)^{1/2} \qquad (4.82)$$

a result which can also be established more rigorously via a striation-thickness analysis.

There are two approaches to the evaluation of over-all mixing in a melt flow. In the first, equation 4.82 can be integrated along the path taken by a particular element of material to obtain both the degree of mixing and residence time as functions of the initial position of the element in the flow channel. While such information emphasises variations of the degree of mixing over the channel, a single parameter for describing the over-all mixing performance of a process is often more useful. Such a parameter is provided by the alternative approach. Applied to a melt flow of the form illustrated in figure 4.2, this involves first evaluating the mean mixing rate over each channel cross-section

$$\frac{d\overline{M}}{dt} = \frac{1}{H} \int_{-H/2}^{+H/2} (4I_2)^{1/2} \, dy \qquad (4.83)$$

This leads to the bulk mean mixing per unit channel length in the direction of flow

$$\overline{M}_x = \frac{d\overline{M}}{dx} = \frac{1}{Q} \int_{-H/2}^{+H/2} (4I_2)^{1/2} \, dy \qquad (4.84)$$

where Q is the volumetric flow rate per unit width, and finally to the total bulk mean mixing

$$\overline{M} = \frac{1}{Q} \int_0^L \int_{-H/2}^{+H/2} (4I_2)^{1/2} \, dy \, dx \qquad (4.85)$$

The value of $\overline{M}$ obtained in this way depends on the particular form of the velocity profiles involved. It should be recalled, however, that in establishing the constitutive equation in section 4.3.1, the melt was assumed to be homogeneous and isotropic. On the other hand, mixing has been defined as an homogenising process. If the assumption of homogeneity is rejected, however, little progress is possible towards obtaining even the simplest solutions, and it must therefore be retained. Difficulties arise, for example, when components with substantially different viscous properties are processed together, unless one component is very predominant.

5

Some Melt Flow Processes

In chapter 4, the fundamental continuum mechanics equations governing the flow
of polymer melts were established. These may now be used to analyse the types of
flows occurring in practical processing operations. In addition to aiding the understanding of observed flow behaviour, such analyses are often capable of providing
rational means of designing processing equipment to meet particular performance
requirements.

In this chapter, analyses of some of the simpler melt flow processes are undertaken. Examples include flows in extrusion dies and between calender rolls and also
in the thin intensely sheared melt films often formed at the surfaces of compacted
solid polymers during melting. Melt flows in extruders and injection-moulding
equipment, which are more complex, are treated in later chapters. Although
attention is confined to internal flows bounded by solid surfaces, this should not be
taken to mean that free surface flows of the type involved in tubular film
production, blow moulding and spinning are not also important. Such flows are in
many respects more difficult to analyse than internal ones, not least because the
positions of the flow boundaries are not known in advance. The interested reader is
referred to the list of further reading for some of the published work on this subject.

An important practical difference between internal and free surface flows is that,
with the latter, it is possible to exert a much greater degree of control while the
process is running. For example, in screw extrusion, the screw speed is the main
process variable, influencing primarily the production rate, while barrel temperature
settings have rather less effect. In tubular film production, however, film dimensions
and mechanical properties are very sensitive to changes in internal air pressure,
cooling rate and haul off speed. In other words, while the performance of internal
flow processes is very dependent on hardware design, which can be assisted by flow
analysis, that of free surface flows is influenced more by process control.

5.1 Some Simple Extrusion Dies

From the earlier descriptions of extrusion processes, it is clear that melt flows in
many types of die are essentially unidirectional and occur in channels of
comparatively simple shape, often of circular, flat or annular cross section. Such

geometries and flows are relatively easy to treat analytically and, provided isothermal flow conditions are assumed, formulae relating flow rates and pressure differences can be derived for use in die design. The isothermal flow assumption was discussed in some detail in section 4.5, and a test for its validity was established, based on the Griffith number for the flow. This criterion is often satisfied quite well in extrusion dies. If it is not, then developing flow analyses of the types described in later chapters for flow in extruders and injection moulding equipment should ideally be used.

5.1.1 Dies of Circular Cross-section

Consider the steady flow along a uniform tube of diameter D, shown in figure 5.1. Cylindrical polar coordinates r and z are appropriate for describing this axisymmetric flow, in which there are no variations with respect to the third angular coordinate, θ. Assuming the lubrication approximation described in section 4.6 to be applicable, the only variable to change with axial position is the pressure, p, and equation 4.15 for equilibrium in the direction of flow reduces to

$$\frac{1}{r}\frac{\mathrm{d}}{\mathrm{d}r}(r\tau_{zr}) = \frac{\mathrm{d}p}{\mathrm{d}z} \equiv P_z \tag{5.1}$$

where P_z is the axial pressure gradient. Since axial symmetry ensures that the shear stress is zero at the axis, this equation may be integrated to give the stress distribution

$$\tau_{zr} = \tfrac{1}{2}rP_z \tag{5.2}$$

which agrees with equation 3.10 derived from a much more rudimentary analysis of flow in a rheometer capillary. According to equations 4.4 and 4.5, the only nonzero rates of deformation are the complementary e_{rz} and e_{zr} defined by equations 4.32. Hence, using the generalised power-law form given by equation 4.38 of the stokesian constitutive equation 4.18

$$\tau_{zr} = \mu_0 \left| \frac{\sqrt{(4I_2)}}{\gamma_0} \right|^{n-1} \frac{\mathrm{d}w}{\mathrm{d}r}, I_2 = \frac{1}{4}\left(\frac{\mathrm{d}w}{\mathrm{d}r}\right)^2 \tag{5.3}$$

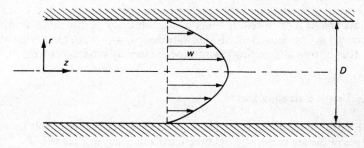

Figure 5.1   *Melt flow along a uniform circular tube*

where, by virtue of the isothermal flow assumption, the reference viscosity, μ_0, is independent of temperature. Combining equations 5.2 and 5.3

$$\mu_0 \left| \frac{1}{\gamma_0} \frac{dw}{dr} \right|^{n-1} \frac{dw}{dr} = \frac{rP_z}{2} \tag{5.4}$$

An analytical difficulty inherent in the empirical power law for describing viscosity variations is now apparent. The result of integrating this equation to find the velocity profile depends on the sign of the velocity gradient, and therefore of the shear stress. In this particular case, however, the sign does not change and is negative: in order for melt to flow in the positive z direction as shown, the pressure gradient must be negative, and τ_{zr} takes the same sign as P_z according to equation 5.2. Alternatively, considering the velocity profile sketched in figure 5.1, w decreases continuously from the axis to the wall of the tube, implying that the velocity gradient is always negative. Therefore, equation 5.4 can be rearranged as

$$\frac{dw}{dr} = -\left[\frac{\gamma_0^{n-1}(-P_z)}{\mu_0} \frac{}{2} \right]^{1/n} r^{1/n} = -C_1 r^{1/n} \tag{5.5}$$

where C_1 is merely a convenient shorthand for the constant terms on the right-hand side of the equation. Integration yields for the velocity profile

$$w = -\frac{C_1 r^{1/n+1}}{\left(\frac{1}{n} + 1 \right)} + A \tag{5.6}$$

where A is an integration constant. Introducing the no-slip boundary condition discussed in section 4.4, $w = 0$ at $r = D/2$, and

$$w = \frac{C_1}{\left(\frac{1}{n} + 1 \right)} \left[\left(\frac{D}{2} \right)^{1/n+1} - r^{1/n+1} \right] \tag{5.7}$$

Figure 5.2 illustrates the shapes of velocity profiles predicted by this equation for three values of the power-law index, with the same volumetric flow rate in each

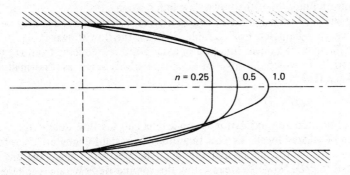

Figure 5.2 *Velocity profiles for non-newtonian flow in circular tubes or between flat stationary boundaries*

case. Note how the parabolic shape associated with newtonian fluids, given by $n = 1$, becomes increasingly flat in the middle of the tube as the melt becomes more pseudoplastic.

The volumetric flow rate along the tube can be obtained by integrating the velocity profile as

$$Q = \int_0^{D/2} 2\pi r w \, dr = \frac{\pi C_1}{\left(\dfrac{1}{n} + 3\right)} \left(\frac{D}{2}\right)^{1/n+3} \tag{5.8}$$

For a newtonian fluid, this result reduces to the form displayed in equation 3.14. Note that, because, according to equation 5.5, C_1 depends on the n^{-1}th power of the pressure gradient, there is a non-linear relationship between flow rate and pressure gradient of the form

$$P_z \propto Q^n \tag{5.9}$$

This relationship is true for isothermal melt flow along a channel of any shape, and represents a fundamental feature of polymer melt processes: an increase in flow rate is accompanied by a much less than proportional increase in the associated pressures.

Now, equation 5.8, together with equation 5.5 for the definition of C_1, provides a rather cumbersome way of relating flow rate to pressure gradient for flow in a die of circular cross-section. It is convenient to introduce a mean shear stress, defined as in section 4.5 in terms of a mean flow velocity, $\overline{U}$, and mean shear rate, $\overline{\gamma}$, as follows

$$\overline{U} = \frac{4Q}{\pi D^2} \,, \overline{\gamma} = \frac{\overline{U}}{D} \,, \overline{\tau} = \mu_0 \left|\frac{\overline{\gamma}}{\gamma_0}\right|^{n-1} \overline{\gamma} \tag{5.10}$$

Hence, using equations 5.8 and 5.5, a dimensionless pressure gradient can be defined as

$$\pi_P = \frac{P_z D}{\overline{\tau}} = -2^{n+2} \left(\frac{3n+1}{n}\right)^n \tag{5.11}$$

which is now a function of the power-law index alone.

Having analysed flow in a die of constant cross-section, the result may be applied to one of varying diameter. For example, consider the flow shown in figure 5.3, where the diameter is a linear function of axial position, reducing from D_1 to D_2 over a length L. Provided the angle of taper is relatively small, as it normally is in practice, such that

$$D_1 - D_2 \ll L \tag{5.12}$$

then the lubrication approximation is applicable, and the flow at any die cross section can be treated locally as flow in a tube of constant diameter. Therefore, if at a distance z along the die the diameter is D, the pressure gradient there is proportional to $\overline{\tau}/D$. Now, for steady flow the volumetric flow rate is independent of axial position, a statement that is essentially an integral form of continuity equation 4.8, and from equations 5.10 the mean shear stress is proportional to

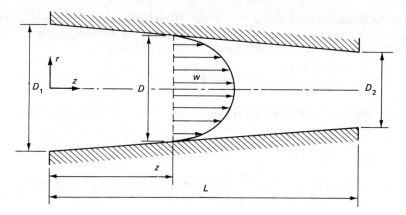

Figure 5.3   *Flow along a tapered circular tube*

D^{-3n}. Hence, the pressure gradient at the typical section of diameter D is

$$P_z = P_{z1} \left(\frac{D}{D_1} \right)^{-1-3n} \tag{5.13}$$

where P_{z1} is the gradient at the initial section. The over-all pressure difference my therefore be determined as

$$p_1 - p_2 = -\int_0^L P_z \, dz = -P_{z1} \int_{D_1}^{D_2} \left(\frac{D}{D_1} \right)^{-1-3n} \left(\frac{dz}{dD} \right) dD \tag{5.14}$$

For a linear variation of D with z

$$D = D_1 + (D_2 - D_1) \frac{z}{L} \tag{5.15}$$

and

$$
\begin{aligned}
p_1 - p_2 &= \frac{P_{z1} L}{(D_1 - D_2)} \int_{D_1}^{D_2} \left(\frac{D}{D_1} \right)^{-1-3n} dD \\
&= \frac{P_{z1} L D_1}{3n(D_1 - D_2)} \left[\left(\frac{D_2}{D_1} \right)^{-3n} - 1 \right]
\end{aligned} \tag{5.16}
$$

The following practical example illustrates the use of the above formulae for pressure gradient and pressure difference. A typical low-density polyethylene with melt flow index 0.2 (see section 3.3.2) has the following properties at a processing temperature of 250 °C

$$\mu_0 = 16.0 \text{ kN s/m}^2 \text{ at } \gamma_0 = 1 \text{ s}^{-1}, n = 0.30, b = 0.01 \text{ °C}^{-1}$$

$$\rho = 750 \text{ kg/m}^3, C_p = 3400 \text{ J/kg °C}, k = 0.30 \text{ W/m °C}$$

Note the use of unit shear rate as a convenient reference value for viscosity data. Suppose this material flows at a rate $Q = 7.5 \times 10^{-7} \text{ m}^3/\text{s}$ in a linearly tapered circular die whose dimensions are $D_1 = 4$ mm, $D_2 = 2$ mm and $L = 50$ mm. Clearly, the geometric condition stated in equation 5.12 is satisfied. From equation 5.11, the the dimensionless pressure gradient for the present material is

$$\pi_P = \frac{P_z D}{\bar{\tau}} = -2^{2.3} \left(\frac{1.9}{0.3}\right)^{0.3} = -8.57$$

Also, at the die inlet, the mean velocity, shear rate, viscosity and shear stress are

$$\bar{U}_1 = \frac{4Q}{\pi D_1^{\,2}} = 59.7 \text{ mm/s}, \; \bar{\gamma}_1 = \frac{\bar{U}_1}{D_1} = 14.9 \text{ s}^{-1}$$

$$\bar{\mu}_1 = \mu_0 \left|\frac{\bar{\gamma}_1}{\gamma_0}\right|^{n-1} = 2.41 \text{ kN s/m}^2, \; \bar{\tau}_1 = \bar{\mu}_1 \bar{\gamma}_1 = 36.0 \text{ kN/m}^2$$

Referring back to the dimensionless parameters established in section 4.5, the characteristic Reynolds number, Peclet number, Griffith number and Graetz number for the flow may be defined in terms of conditions at the die entry as

$$Re = \frac{\rho \bar{U}_1 D_1}{\bar{\mu}_1} = \frac{750 \times 59.7 \times 10^{-3} \times 4 \times 10^{-3}}{2.41 \times 10^3} = 7.43 \times 10^{-5}$$

$$Pe = \frac{\rho C_p \bar{U}_1 D_1}{k} = \frac{750 \times 3400 \times 59.7 \times 10^{-3} \times 4 \times 10^{-3}}{0.30} = 2030$$

$$G = \frac{b \bar{\tau}_1 \bar{\gamma}_1 D_1^{\,2}}{k} = \frac{0.01 \times 36.0 \times 10^3 \times 14.9 \times (4 \times 10^{-3})^2}{0.30} = 0.286$$

$$Gz = Pe \frac{D_1}{L} = 2030 \times \frac{4}{50} = 162$$

The typically small Reynolds number confirms that inertia effects are indeed negligible. As the Griffith number is significantly less than unity, the isothermal assumption is reasonable. With such large values of both Pe and Gz, thermal convection is the dominant mode of heat transfer, although in view of the isothermal condition the temperatures developed do not affect the relationship between flow rate and pressures.

Having evaluated the dimensionless pressure gradient, the actual gradient at the inlet may be found as

$$P_{z1} = \frac{\pi_P \bar{\tau}_1}{D_1} = -\frac{8.57 \times 36.0}{4} = -77.1 \text{ MN/m}^3$$

and, from equation 5.16, the over-all pressure difference is

$$p_1 - p_2 = \frac{77.1 \times 50 \times 10^{-3} \times 4}{0.9 \times (4-2)} \left[\left(\frac{2}{4}\right)^{-0.9} - 1\right] = 7.42 \text{ MN/m}^2$$

For comparison, if the die had been of constant diameter D, then

$$p_1 - p_2 = -P_{z1} L = 77.1 \times 50 \times 10^{-3} = 3.85 \text{ MN/m}^2$$

5.1.2 Flat Slit Dies

The flow in a number of types of extrusion die can be treated as flow between parallel, or nearly parallel, flat stationary boundaries of infinite extent. Appropriate channel geometry and coordinates are defined in figure 4.2. Assuming the lubrication approximation to be applicable, equilibrium equation 4.73 can be integrated to give

$$\tau_{xy} = yP_x \tag{5.17}$$

by virtue of the symmetry condition $\tau_{xy} = 0$ at $y = 0$, the mid-plane of the flow. With e_{xy} and e_{yx} being the only non-zero rates of deformation, the generalised power-law constitutive equation 4.38 yields

$$\tau_{xy} = \mu_0 \left| \frac{\sqrt{(4I_2)}}{\gamma_0} \right|^{n-1} \frac{du}{dy}, \qquad I_2 = \frac{1}{4}\left(\frac{du}{dy}\right)^2 \tag{5.18}$$

Reference viscosity, μ_0, is independent of temperature in the assumed isothermal flow. Combining equations 5.17 and 5.18

$$\mu_0 \left| \frac{1}{\gamma_0} \frac{du}{dy} \right|^{n-1} \frac{du}{dy} = yP_x \tag{5.19}$$

Although there is a change in sign of the velocity gradient at the mid-plane, the analytical difficulty associated with it can be avoided by considering only one symmetrical half of the flow. For example, in the region $y \geqslant 0$, the velocity gradient is zero or negative, and equation 5.19 can be rearranged as

$$\frac{du}{dy} = -\left[\frac{\gamma_0^{n-1}}{\mu_0}(-P_x) \right]^{1/n} y^{1/n} = -C_2 y^{1/n} \tag{5.20}$$

where C_2 represents the combined constant terms on the right-hand side of the equation. Integration and application of the no-slip boundary condition, $u = 0$ at $y = H/2$, gives the velocity profile

$$u = \frac{C_2}{\left(\dfrac{1}{n}+1\right)} \left[\left(\frac{H}{2}\right)^{1/n+1} - y^{1/n+1} \right] \tag{5.21}$$

Note the close similarity with the velocity profile in a channel of circular cross section given by equation 5.7: the profile shapes illustrated in figure 5.2 hold for flow in a flat slit.

The volumetric flow rate per unit width of the slit is given by

$$Q = 2 \int_0^{H/2} u\, dy = \frac{2C_2}{\left(\dfrac{1}{n}+2\right)} \left(\frac{H}{2}\right)^{1/n+2} \tag{5.22}$$

which, for a newtonian fluid of viscosity μ, becomes

$$Q = -\frac{P_x H^3}{12\mu} \tag{5.23}$$

As in the case of flow in a circular channel, the relationship between pressure gradient and flow rate is of the form given by equation 5.9. It is again convenient to define a mean shear stress in terms of a mean velocity and mean shear rate

$$\overline{U} = \frac{Q}{H}, \qquad \overline{\gamma} = \frac{\overline{U}}{H}, \qquad \overline{\tau} = \mu_0 \left| \frac{\overline{\gamma}}{\gamma_0} \right|^{n-1} \overline{\gamma} \qquad (5.24)$$

dimensionless pressure gradient can be defined with the aid of equations 5.22 and 5.20 as

$$\pi_P = \frac{P_x H}{\overline{\tau}} = -2^{2n+1} \left(\frac{2n+1}{2n} \right)^n \qquad (5.25)$$

Having analysed flow in a channel of constant depth, the result may be applied to one of varying depth. For example, consider the case of a linear depth variation from H_1 to H_2 over a distance L in the direction of flow. Provided the angle of taper is small, the lubrication approximation is applicable. Following the argument developed in the last subsection, the pressure gradient at the typical channel section of depth H is

$$P_x = P_{x1} \left(\frac{H}{H_1} \right)^{-1-2n} \qquad (5.26)$$

where P_{x1} is the gradient at the channel inlet. The over-all pressure difference is

$$p_1 - p_2 = -\int_0^L P_x \, dx = -P_{x1} \int_{H_1}^{H_2} \left(\frac{H}{H_1} \right)^{-1-2n} \left(\frac{dx}{dH} \right) dH \qquad (5.27)$$

and for a linear variation of H with x

$$H = H_1 + (H_2 - H_1) \frac{x}{L} \qquad (5.28)$$

giving

$$p_1 - p_2 = -\frac{P_{x1} L H_1}{2n(H_1 - H_2)} \left[\left(\frac{H_2}{H_1} \right)^{-2n} - 1 \right] \qquad (5.29)$$

Other forms of depth variation can of course be accommodated, although the pressure gradient integration in equation 5.27 may be less straightforward, and in many cases can only be carried out numerically.

Although the present flat-channel flows are assumed to be isothermal in the sense that velocity and temperature profiles are not coupled via the dependence of viscosity on temperature, it is still possible to determine the temperature profile. If the lubrication approximation is applied to temperatures — which is a much less reasonable procedure than in the case of velocities, for the reasons discussed in section 4.6 — the simplified energy equation 4.76 may be expressed in the form

$$\frac{d^2 T}{dy^2} = -\frac{\mu}{k} \left(\frac{du}{dy} \right)^2 \qquad (5.30)$$

Introducing the power-law definition of viscosity, μ, implied in equation 5.18

$$\frac{d^2 T}{dy^2} = -\frac{\mu_0}{k \gamma_0^{n-1}} \left| \frac{du}{dy} \right|^{n+1} \qquad (5.31)$$

which, using equation 5.20 for the velocity gradient and equation 5.25 for the relationship between π_P and P_x, becomes

$$\frac{d^2T}{dy^2} = -\frac{\overline{\tau\gamma}}{k}\left(-\frac{\pi_P y}{H}\right)^{1/n+1} = -C_3 y^{1/n+1} \qquad (5.32)$$

for $y \geqslant 0$, where C_3 represents the combined constant terms. Hence, assuming the temperature profile to be symmetrical about the mid plane of the flow, for the upper half

$$\frac{dT}{dy} = -\frac{C_3 y^{1/n+2}}{\left(\dfrac{1}{n}+2\right)} \qquad (5.33)$$

and, assuming controlled boundary temperatures, T_b, with good thermal contact at the solid–melt interfaces as discussed in section 4.4

$$T - T_b = \frac{C_3}{\left(\dfrac{1}{n}+2\right)\left(\dfrac{1}{n}+3\right)}\left[\left(\frac{H}{2}\right)^{1/n+3} - y^{1/n+3}\right]$$

$$= \frac{\overline{\tau\gamma}H^2(-\pi_P)^{1/n+1}}{k\left(\dfrac{1}{n}+2\right)\left(\dfrac{1}{n}+3\right)}\left[\left(\frac{1}{2}\right)^{1/n+3} - \left(\frac{y}{H}\right)^{1/n+3}\right] \qquad (5.34)$$

The following practical example illustrates the use of the above formula for flow and temperature rise. Consider a low-density polyethylene having the same properties as given in the example in the last subsection, flowing at a volumetric rate per unit width $Q = 4.8 \times 10^{-4}$ m^2/s in a wide flat die of depth $H = 6$ mm and length $L = 250$ mm. The mean velocity, shear rate, viscosity and shear stress are

$$\overline{U} = \frac{Q}{H} = 80 \text{ mm/s}, \ \overline{\gamma} = \frac{\overline{U}}{H} = 13.3 \text{ s}^{-1}$$

$$\overline{\mu} = \mu_0 \left|\frac{\overline{\gamma}}{\gamma_0}\right|^{n-1} = 2.61 \text{ kN s/m}^2, \ \overline{\tau} = \overline{\mu}\overline{\gamma} = 34.8 \text{ kN/m}^2$$

The characteristic Reynolds, Peclet, Griffith and Graetz numbers for the flow are

$$Re = \frac{\rho\overline{U}H}{\overline{\mu}} = \frac{750 \times 80 \times 10^{-3} \times 6 \times 10^{-3}}{2.61 \times 10^3} = 1.38 \times 10^{-4}$$

$$Pe = \frac{\rho C_p \overline{U}H}{k} = \frac{750 \times 3400 \times 80 \times 10^{-3} \times 6 \times 10^{-3}}{0.30} = 4080$$

$$G = \frac{b\overline{\tau\gamma}H^2}{k} = \frac{0.01 \times 34.8 \times 10^3 \times 13.3 \times (6 \times 10^{-3})^2}{0.30} = 0.555$$

$$Gz = Pe\frac{H}{L} = 4080 \times \frac{6}{250} = 97.9$$

While inertia effects are, as usual, negligible, the size of G implies that the isothermal assumption is only marginally justified. Thermal convection is the dominant mode of heat transfer.

From equation 5.25, the dimensionless pressure gradient is

$$\pi_P = \frac{P_x H}{\bar{\tau}} = -2^{1.6} \left(\frac{1.6}{0.6}\right)^{0.3} = -4.07$$

giving an actual pressure gradient of

$$P_x = \frac{\pi_P \bar{\tau}}{H} = -\frac{4.07 \times 34.8}{6} = -23.6 \text{ MN/m}^3$$

and an over-all pressure difference

$$p_1 - p_2 = -P_x L = 23.6 \times 0.25 = 5.9 \text{ MN/m}^2$$

Also, from equation 5.34, the maximum temperature in thermally fully developed flow is at the mid-plane, $y = 0$, where

$$T - T_b = \frac{34.8 \times 10^3 \times 13.3 \times (6 \times 10^{-3})^2 \times 4.07^{4.33}}{0.3 \times 5.33 \times 6.33} \left(\frac{1}{2}\right)^{6.33}$$

$$= 8.9\ ^\circ C$$

That the isothermal assumption is only marginally justified is confirmed by the fact that, with a temperature coefficient of viscosity $b = 0.01\ ^\circ C^{-1}$, a temperature rise of this magnitude causes about a 9 per cent reduction in viscosity. On the other hand, fully developed flow conditions are only achieved after a length of flow of the order of that given by equation 4.69 as

$$L' = \frac{\rho C_p \bar{U} H^2}{10k} = \frac{LGz}{10} = 2.45 \text{ m}$$

which is nearly 10 times the actual length of flow. In other words, temperature-profile development is far from complete at the end of the flow, and the maximum temperature rises generated are only of the order of $1-2\ ^\circ C$.

5.1.3 Annular Dies

The flow in some types of extrusion die can be treated as flow between concentric cylindrical surfaces forming an annulus. Consider such a flow, whose geometry and coordinates are shown in figure 5.4. Applying the lubrication approximation in the usual way, the flow may be assumed to be locally fully developed and also isothermal, and equation 5.1 governs equilibrium in the direction of flow. Integration yields

$$\tau_{zr} = \tfrac{1}{2} r P_z + \frac{B}{r} \tag{5.35}$$

where P_z is the axial pressure gradient, B is an integration constant and the shear

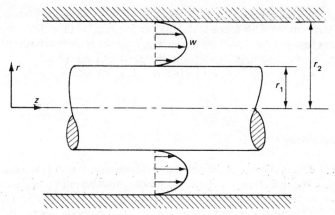

Figure 5.4   *Flow along an annulus*

stress is related to the velocity gradient, dw/dr, by equation 5.3. Hence

$$\mu_0 \left| \frac{1}{\gamma_0} \frac{dw}{dr} \right|^{n-1} \frac{dw}{dr} = \frac{rP_z}{2} + \frac{B}{r} \qquad (5.36)$$

which may be rearranged for the velocity gradient as

$$\frac{dw}{dr} = \text{sgn}\,(E)\,|\,E\,|^{1/n}, \qquad E = \frac{\gamma_0^{\,n-1}}{\mu_0} \left(\frac{rP_z}{2} + \frac{B}{r} \right) \qquad (5.37)$$

where 'sgn' means 'the sign of'.

The analytical difficulty associated with the empirical power-law constitutive equation is now encountered. The form of integral, and associated integration constant, for velocity w from equation 5.37 depends on the sign of the velocity gradient and shear stress. These two expressions must, of course, be matched at the stress neutral surface, where the shear stress is zero. The exact position of this surface is not known in advance but, for annular flow between stationary boundaries, it must lie in the range $r_1 < r < r_2$. In the previous examples of flow in circular and flat dies, the stress neutral position was fixed by symmetry considerations, and the analysis confined to a region of flow where the stress did not change sign. In general, for a power-law melt flow, if the stress neutral surface is at some unknown position within the flow, then an analytical solution for the velocity profile is not possible. With an arbitrary value of n, equation 5.37 can only be solved numerically to satisfy the no-slip boundary conditions.

If the annulus is such that the radial gap is small compared with the mean radius, then it is reasonable to unroll the flow channel and treat it as a flat slit with a depth of $(r_2 - r_1)$, and a width equal to the mean circumference of the annulus, namely $\pi(r_1 + r_2)$. Therefore, adapting equations 5.24 and 5.25

$$\bar{U} = \frac{Q}{\pi(r_2^{\,2} - r_1^{\,2})}, \; \bar{\gamma} = \frac{\bar{U}}{(r_2 - r_1)}, \; \bar{\tau} = \mu_0 \left| \frac{\bar{\gamma}}{\gamma_0} \right|^{n-1} \bar{\gamma} \qquad (5.38)$$

$$\pi_P = \frac{P_z(r_2 - r_1)}{\bar{\tau}} = -2^{2n+1} \left(\frac{2n+1}{2n} \right)^n \qquad (5.39)$$

where P_z is the axial pressure gradient, and Q the total volumetric flow rate in the annulus. Comparisons with numerical solutions for annular flow show this approximate result to be remarkably accurate, even for relatively large values of the ratio $K = r_2/r_1$. For example, the error in the pressure gradient is less than 1 per cent when $K = 2$, and less than 2 per cent when $K = 3$, for typical values of the power-law index.

5.1.4 Wire-covering Dies

A somewhat different form of the annular-flow problem is to be found in wire-covering dies, the main distinction being that the inner boundary in figure 5.4, which is the wire itself, moves at a constant speed, say V. The die that forms the outer boundary is usually tapered to reduce the cross sectional area of flow gradually, but not so rapidly as to invalidate the lubrication approximation in terms of geometry changes. Equation 5.37 still defines the velocity gradient, although the no-slip velocity boundary conditions are now $w = V$ at $r = r_1$ and $w = 0$ at $r = r_2$. Once again, the velocity profile cannot, in general, be obtained analytically.

One special case is that of isothermal flow with no pressure gradient, when equation 5.37 becomes

$$\frac{dw}{dr} = -\left[\frac{\gamma_0{}^{n-1}}{\mu_0} \frac{(-B)}{r} \right]^{1/n} = -C_4 r^{-1/n} \tag{5.40}$$

from which

$$w = V \frac{r_2{}^{1-1/n} - r^{1-1/n\cdot}}{r_2{}^{1-1/n} - r_1{}^{1-1/n}} \tag{5.41}$$

A dimensionless volumetric flow rate can be defined as the ratio between the actual flow rate and the rate that would be obtained if all the melt moved with the speed of the wire

$$\pi_Q = \frac{Q}{\pi(r_2{}^2 - r_1{}^2)V} , \qquad Q = \int_{r_1}^{r_2} 2\pi r w \, dr \tag{5.42}$$

For the velocity profile defined by equation 5.41

$$\pi_Q = \frac{K^{1-1/n}}{K^{1-1/n} - 1} - \frac{2(K^{3-1/n} - 1)}{(3 - 1/n)(K^{1-1/n} - 1)(K^2 - 1)} \tag{5.43}$$

where $K = r_2/r_1$, the cases $n = 1$ and $1/3$ requiring special treatment. In a practical problem, if r_1, r_2, V and n are given, this result allows the dimensionless flow rate, and hence the actual flow rate, to be determined.

Another special case is that of isothermal newtonian flow, when equation 5.37 becomes

$$\frac{dw}{dr} = \frac{1}{\mu} \left(\frac{rP_z}{2} + \frac{B}{r} \right) \tag{5.44}$$

Repeated integration of this equation yields the flow rate, which may be expressed in the dimensionless form defined in equation 5.42 as

$$\pi_Q = \frac{1}{2 \ln (K)} - \frac{1}{K^2 - 1} - \frac{\pi_P}{8(K-1)^2}\left[K^2 + 1 - \frac{K^2 - 1}{\ln (K)}\right] \quad (5.45)$$

where π_P is the dimensionless pressure gradient

$$\pi_P = \frac{P_z(r_2 - r_1)}{\bar{\tau}}, \qquad \bar{\tau} = \mu\bar{\gamma} = \frac{\mu V}{(r_2 - r_1)} \quad (5.46)$$

Although attention is confined here to flow rates and pressure gradients in wire covering dies, other variables are also of considerable importance in practice. These include the tension gradient in the wire, which is proportional to the shear stress applied to its surface, and the shear stress at the die wall, which, if large, can lead to slip and melt fracture, as discussed in sections 4.4, 3.3.3 and 3.4. Such information is readily obtainable from the flow analysis. References to more detailed treatments of the wire covering process, including numerical solutions to the flow equations, are given in the list of further reading.

Although the flow equations cannot usually be solved analytically, this in no way prevents the nature of melt flow in wire-covering dies being examined in terms of the usual dimensionless parameters. Consider a practical example involving a wire-covering grade of low-density polyethylene having the following melt properties

$$\mu_0 = 2.0 \text{ kN s/m}^2 \text{ at } \gamma_0 = 1 \text{ s}^{-1}, n = 0.48, b = 0.01 \,^\circ\text{C}^{-1}$$

$$\rho = 750 \text{ kg/m}^3, C_p = 2500 \text{ J/kg }^\circ\text{C}, k = 0.30 \text{ W/m }^\circ\text{C}$$

The wire is of radius $r_1 = 0.5$ mm and moves at $V = 20$ m/s. The characteristic die dimensions can be taken as $r_2 = 1$ mm, axial length $L = 25$ mm. The characteristic Reynolds, Peclet, Griffith and Graetz numbers are therefore

$$Re = \frac{\rho V(r_2 - r_1)}{\bar{\mu}} = 0.927$$

$$Pe = \frac{\rho C_p V(r_2 - r_1)}{k} = 6.25 \times 10^4$$

$$G = \frac{b\bar{\tau}\bar{\gamma}(r_2 - r_1)^2}{k} = 108$$

$$Gz = Pe\frac{(r_2 - r_1)}{L} = 1250$$

where the mean shear rate is $\bar{\gamma} = V/(r_2 - r_1) = 4 \times 10^4 \text{ s}^{-1}$. High-speed wire covering operations provide a rare example in polymer processing of a flow situation where inertia effects can become significant: Re in this case is of order 1. Although the very large Griffith number implies that the isothermal assumption would be extremely inappropriate for thermally fully developed flow, the even larger sizes of both Pe and Gz ensure that such a condition is not even approached. In other words, over the short length of the die, the melt temperature rises as a result of convected heat are insufficient to affect melt viscosities significantly. An alternative, but equivalent, argument for retaining the isothermal flow assumption is similar to that outlined in section 3.3.3 for capillary rheometer flow, in which a mean melt-temperature rise is calculated from the work done on the melt by, in this case, the motion of the wire.

5.2 Narrow Channel Flows in Dies and Crossheads

The polymer melt flows in simple extrusion dies treated in section 5.1 were all essentially one dimensional in that there was only one non-zero velocity component in each case. While more general two- and three-dimensional flows are considered later in connection with extruder screw channels, there are a number of types of dies and crossheads in which the flows must be treated as being two dimensional. This is the case when the flow channel is narrow in the sense that the dimension normal to the plane of flow, namely the channel depth, is small compared to the other two dimensions, and only varies slowly over the region of interest. Practical examples include flat and tubular film dies, wire and cable covering crossheads and pipe dies.

In such narrow channel flows, the lubrication approximation is applicable to velocities, and it is also often reasonable to assume isothermal conditions, thus avoiding the need to consider temperature profiles. Consider the flow illustrated in figure 5.5, in which the local resultant direction of motion is s, at some arbitrary direction to the x and z axes in the local plane of the boundaries. The third coordinate, y, is normal to this plane. Let Q_s be the volumetric flow rate in the s direction per unit width of channel normal to this direction. As the flow is locally one dimensional between flat parallel boundaries, the pressure gradient in the s direction is

$$\frac{\partial p}{\partial s} = \frac{\pi_P \bar{\mu} Q_s}{H^3} \tag{5.47}$$

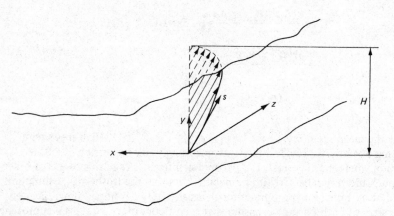

Figure 5.5   *Geometry and coordinates for melt flow in a narrow channel*

where π_P is the function of the power-law index given by equation 5.25, and $\bar{\mu}$ is the viscosity evaluated at the local mean shear rate, Q_s/H^2

$$\bar{\mu} = \mu_0 \left(\frac{Q_s}{\gamma_0 H^2}\right)^{n-1} \tag{5.48}$$

Although the shape of the resultant velocity profile does not change in a narrow channel flow, its direction may vary slowly with position, that is, with coordinates x and z in figure 5.5. Let Q_x and Q_z be the component flow rates in the x and z directions, per unit width of channel in the z and x directions respectively. Hence

$$Q_s^2 = Q_x^2 + Q_z^2 \tag{5.49}$$

With no velocity component in the y direction, the differential continuity equation 4.8 is

$$\frac{\partial u}{\partial x} + \frac{\partial w}{\partial z} = 0 \tag{5.50}$$

which, with $u = u(y)$ and $w = w(y)$, may be integrated with respect to y to give

$$\frac{\partial}{\partial x}\left(\int_0^H u\,dy\right) + \frac{\partial}{\partial z}\left(\int_0^H w\,dy\right) = 0$$

or

$$\frac{\partial Q_x}{\partial x} + \frac{\partial Q_z}{\partial z} = 0 \tag{5.51}$$

This equation is automatically satisfied by the following stream function, $\psi(x, z)$

$$Q_x = \frac{\partial \psi}{\partial z}, \qquad Q_z = -\frac{\partial \psi}{\partial x} \tag{5.52}$$

Now, using equation 5.47, the pressure gradients in the x and z directions are given by

$$\frac{\partial p}{\partial x} = \frac{\pi_P \bar{\mu} Q_x}{H^3}, \qquad \frac{\partial p}{\partial z} = \frac{\pi_P \bar{\mu} Q_z}{H^3} \tag{5.53}$$

From the fact that the pressure must satisfy the mathematical identity

$$\frac{\partial}{\partial z}\left(\frac{\partial p}{\partial x}\right) - \frac{\partial}{\partial x}\left(\frac{\partial p}{\partial z}\right) = 0 \tag{5.54}$$

can be derived the result

$$\frac{\partial}{\partial x}\left(\frac{\bar{\mu}}{H^3}\frac{\partial \psi}{\partial x}\right) + \frac{\partial}{\partial z}\left(\frac{\bar{\mu}}{H^3}\frac{\partial \psi}{\partial z}\right) = 0 \tag{5.55}$$

Note that $\bar{\mu}$ is still given in terms of Q_s by equation 5.48, and from equation 5.49

$$Q_s^2 = \left(\frac{\partial \psi}{\partial x}\right)^2 + \left(\frac{\partial \psi}{\partial z}\right)^2 \tag{5.56}$$

Provided equation 5.55 can be solved subject to the appropriate conditions at the edges of the region of interest in the x, z plane, the distribution of flow within the narrow channel can be determined from the resulting function $\psi(x, z)$. For example, flow occurs along streamlines, which can be plotted as lines of constant ψ. Also, local rates of flow can be found from the first derivatives of ψ, according to equations 5.52. Equation 5.55 cannot usually be solved analytically, however, and numerical methods must be employed. The equation is of the quasiharmonic type, and its mathematical behaviour is similar to that of the harmonic partial differential equation

$$\frac{\partial^2 \psi}{\partial x^2} + \frac{\partial^2 \psi}{\partial z^2} = 0 \tag{5.57}$$

In view of the irregular shapes of the regions of interest often associated with narrow channel flows, finite element numerical methods of solution are usually to be preferred to other techniques, such as finite difference methods. A suitable finite-element method for the present type of problem is outlined in appendix A. Two practical examples — of flow past the spiders in a pipe die and flow in the deflector of a cable-covering crosshead — are described in the following subsections.

5.2.1 Flow Past Spiders in a Pipe Die

An important application of narrow channel flow analysis is to the determination of the thickness distribution induced in an extrudate by either unsymmetrical flow geometry or obstructions in the flow. For example, in the pipe die illustrated in figure 2.3, the spiders supporting the central mandrel must obstruct the flow to some extent, despite their streamlined shapes. In order to examine this effect quantitatively, the flow in the vicinity of the spiders can be treated as flow in the narrow annular channel unrolled on to the mid-plane of the flow, as shown in figure 5.6a. The coordinates x and z are chosen to be in the circumferential and axial directions of the die, respectively, the origin for x being the centre line of one of the spiders. There are four spiders in this particular example.

The flow is symmetrical about the spider centre lines, and the axial lines midway between the spiders, such as BB'. Assuming that the flow inlet and outlet are the same distances upstream and downstream of the spiders and are subject to the same boundary conditions, and also that the channel depth distribution is symmetrical about the line EF, then the flow is also symmetrical about EF. Therefore, only one sixteenth part of the total flow channel need be analysed. One such part is the shaded region ABCD, which is shown in greater detail in figure 5.6b. This is now the solution domain for equation 5.55. Note that, had the flow not been symmetrical about EF, the larger region ABB'A' would have to be analysed. Both the distribution of channel depth, H, over the region chosen, and the boundary conditions at its edges must be specified before a solution can be attempted. For example, the channel depth might be constant.

As a result either of the assumed symmetry, or of the solid boundary formed by the spider, there is no flow across the boundary joining A and D or that joining B and C. In other words, these boundaries are streamlines, and ψ is constant along each of them. A zero value can be assigned to ψ on boundary AD. Now, since

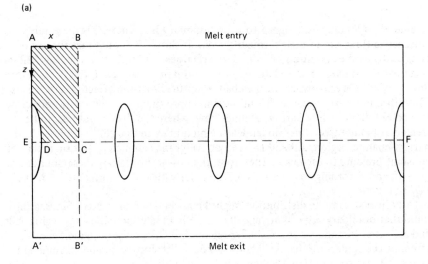

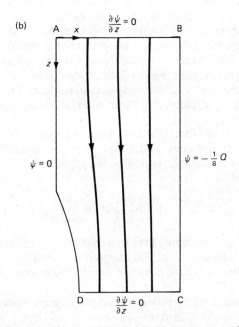

Figure 5.6 *Melt flow past pipe die spiders: (a) the narrow annular channel unrolled; (b) the solution domain and stream function boundary conditions*

equations 5.52 imply that

$$\psi_B - \psi_A = - \int_A^B Q_z \, dx \qquad (5.58)$$

a value of $-Q/8$ may be assigned to ψ on boundary BC, where Q is the total volumetric flow rate of melt through the die. It should be noted that the use of constant ψ as the condition on boundary AD does not imply that the flow velocity is zero at the spider surface. Although the no-slip boundary condition is therefore not satisfied, the narrowness of the channel ensures that the effect is very localised. This difficulty arises because the lubrication approximation is not valid very close to a solid boundary in a narrow-channel flow, where the flow should really be treated as being three dimensional. On the melt inlet and outlet boundaries to the narrow-channel region, AB and DC, respectively, the assumed condition is that of constant pressure in each case. Therefore, there is no flow along these boundaries, $Q_x = 0$, and according to equations 5.52 the derivative of ψ normal to each one, $\partial \psi / \partial z$, is zero.

A typical streamline distribution might be as sketched in figure 5.6b. Bearing in mind that boundary AB can be regarded as both an inlet boundary for region ABCD and an outlet boundary for the similar region A'B'CD shown in figure 5.6a, the effect of the spiders on the extrudate thickness distribution can be examined in terms of the distribution of ψ along AB. The local flow rate in the axial direction, which is directly proportional to the local wall thickness in the finished pipe, can be obtained with the aid of equations 5.52 from the local gradient of ψ along AB. Hence, for the spiders to provide negligible obstruction to the melt flow, the streamlines corresponding to constant increments of ψ should be uniformly distributed along this boundary. In practice, this is achieved by making the axial length of the channel, namely the length BC, large compared to the dimensions of the spiders. In addition to wall thickness variations, the division and re-amalgamation of melt flowing past the spiders may cause relatively weak lines along the finished pipe.

The calculated stream function distribution can also be used to determine the pressure distribution, and in particular the over-all pressure difference in the direction of flow. For example, using equations 5.53 and 5.52, the pressure drop from B to C can be obtained as

$$p_B - p_C = - \int_B^C \left(\frac{\partial p}{\partial z} \right) \, dz = \int_B^C \frac{\pi_p \bar{\mu}}{H^3} \left(\frac{\partial \psi}{\partial x} \right) \, dz \qquad (5.59)$$

The same numerical result would have been obtained by integrating along any other path between the inlet and outlet boundaries of region ABCD. The pressure drop over the entire solution domain shown in figure 5.6a is found by simply doubling that between B and C.

It is worth noting that, for polymer melts obeying the power-law constitutive equation, the form of flow distribution in a narrow channel expressed in terms of streamlines depends only on the power-law index. It is not affected by the volumetric flow rate, or indeed by the channel depth, provided this is varied everywhere in the same proportion. To see this important result, consider equation 5.55, which determines the stream-function distribution. If either H or $\bar{\mu}$, which is dependent on the flow rate and H via equation 5.48, are multiplied by factors that are constant throughout the region of interest, then these factors, or some power-law functions of them, could be eliminated from the equation. Although the form of the stream-function distribution is therefore not affected, the actual numerical values of ψ at particular points in the flow are scaled by the same amount, as are the pressures and pressure differences. In practical terms, this means that a flow channel design, once established for a particular material, should give similar extrudate thickness distributions at all flow rates and processing temperatures. It should only need to be modified if a material with a significantly different power-law index is to be extruded. In terms of the present problem of flow past pipe die spiders, the more non-newtonian the melt, the greater the influence of the spiders on the extrudate thickness distribution.

Another application for narrow channel analysis in pipe dies is to examine the effect of eccentricity of the outer die with respect to the mandrel after the spider region. A further illustration is provided by the following example of a crosshead flow.

5.2.2 Flow in a Cable Covering Crosshead

In the production of covered electrical wire and cable, it is usual to extrude melt into a crosshead through which the conductor is drawn at an angle — frequently a right angle — to the axis of the extruder screw. An illustration of such a crosshead is shown in figure 2.5. Particularly in the case of high-voltage cables, it is not uncommon for a single head unit to be used to apply two or even three layers of different materials during one pass of the conductor. For each layer, the problem is therefore to design a system of flow channels within the head such as to accept a side fed supply of melt and distribute it into a tube of uniform thickness, which is then extruded as part of the cable.

Figure 5.7 shows one commonly used form of arrangement for attempting to achieve the desired uniformity. A narrow radial gap between concentric cylindrical and conical surfaces, the outer members of which have been removed for purposes of illustration, serves to distribute the melt. Since the melt tends to take the shortest path from the inlet to the channel exit, this path is deliberately blocked by a heart-shaped area, which fills the radial gap and forces the melt flow to follow longer paths of more uniform length. The cylindrical portion of the component shown is known as the *deflector*, while the subsequent conically tapered portion is termed the *point*. The channel geometry, and therefore the flow, are intended to be symmetrical about the centre line of the heart-shaped blockage.

Figure 5.8 shows the shape of the one half of the deflector-flow channel unrolled on to the mid-plane of the flow. The coordinates x and z are chosen to be in the circumferential and axial directions of the deflector, respectively, the origin for x being the line of symmetry through the flow inlet. Also shown are some triangular finite elements of the type discussed in appendix A. The points A, B, C and D shown at the intersections of the flow boundaries in figure 5.8 are also marked in figure 5.7. Clearly, flow paths between the inlet boundary AB and outlet CD are of reasonably uniform length. It should be noted that the channel depth is often reduced by tapering in the axial direction.

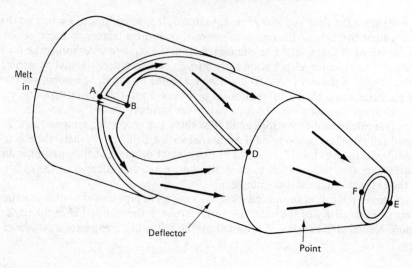

Figure 5.7   *Typical deflector and point used inside a cable covering crosshead*

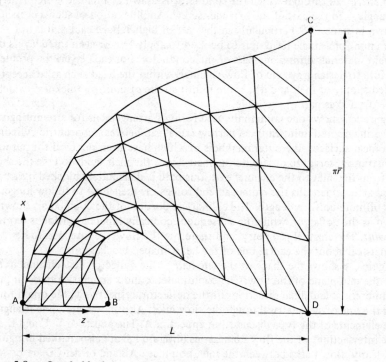

Figure 5.8   *One half of the deflector flow channel unrolled, showing a mesh of finite elements*

The stream function boundary conditions are

$$\psi = 0 \text{ along BD}, \qquad \psi = -\frac{Q}{2} \text{ along AC}$$

$$\frac{\partial \psi}{\partial x} = 0 \text{ along AB}, \qquad \frac{\partial \psi}{\partial z} = 0 \text{ along CD} \qquad (5.60)$$

where Q is the total volumetric rate through both symmetrical halves of the deflector channel. Note that, while the assumption of constant pressure along AB is reasonable, the same assumption along CD deliberately ignores for the moment the presence of the tapered point attached to the deflector.

In order to illustrate the application of the method of narrow-channel analysis, together with the finite-element solution technique described in appendix A, the deflector-channel profile shown in figure 5.8 is considered. The main dimensions are: axial length, L = 170 mm; mean channel radius measured from the common axis of the conductor and deflector, $\bar{r}$ = 45 mm; the channel depth, H, decreases with axial position from 6.2 mm at A to 3.4 mm at boundary CD. A polymer with a power-law index, n = 0.39, and μ_0 = 22.6 kN s/m^2 at γ_0 = 1 s^{-1}, is supplied at a rate Q = 21.4 x 10^{-6} m^3/s to the crosshead. These data are taken from an actual cable making trial. The orders of magnitude calculated for the characteristic Griffith and Graetz numbers for the flow are $G \approx 0.1$ and $Gz \approx 20$, respectively. While the magnitude of Gz implies that thermal convection is the dominant mode of heat transfer, and that thermally fully developed conditions are not reached in the deflector, the size of G ensures that the temperature changes have only small effects on the velocity profiles. The isothermal flow assumption is valid.

Figure 5.9 shows the predicted dimensionless thickness distribution of the polymer leaving the deflector via boundary CD, and the similar boundary for the other half of the deflector. Dimensionless thickness in this context is defined as the local thickness, which is proportional to the local derivative of ψ along the outlet boundary, divided by the average thickness. The distribution shown is the form of variation that would be observed on the finished cable if melt from the deflector were to be applied directly to the conductor. Despite the contouring of its flow channels, the deflector alone clearly gives a poor thickness distribution. In practice, however, the presence of the point after the deflector results in a considerable improvement.

Figure 5.9 also shows the thickness produced by the deflector plus tapered point illustrated in figure 5.7. The axial length of the point is 82 mm, its flow channel is inclined at a mean angle of 23.5° to the axis and the depth changes from 3.4 mm at the interface CD with the deflector to 9.4 mm at the outlet boundary EF. Although the method of narrow-channel analysis has to be somewhat modified to cope with the changing mean radius of the flow channel in the region of the point, the same type of stream-function equation is obtained for the combined flow region ACEFDB. The constant pressure boundary condition is moved forward from boundary CD to EF. Clearly, the addition of the point considerably improves the final thickness distribution on the finished cable. This is because the flow-channel dimensions are such that the pressure difference over the point is substantially greater than that over the deflector. Therefore, the axial symmetry of the tapered

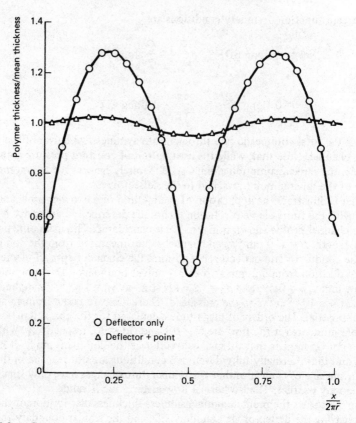

Figure 5.9   *Computed final thickness distributions for deflector alone and combined deflector and point*

point region dominates the lack of symmetry associated with the deflector, to give a reasonably acceptable thickness distribution. If such throttling by the point is not possible, however, the design of the deflector needs to be considerably improved. One possible method for achieving this is discussed in the next section. Predictions of cable layer thickness distributions obtained by the present method have been successfully compared with experimental measurements, as have calculated pressure differences.

5.3 Applications to Die Design

Having established methods for analysing melt flows in various types of extrusion die channels, it is appropriate to consider how these may be applied to the problem of die design. For such a design study, the polymer to be extruded is often specified, together with the required flow rate. The main objective is to establish a flow-channel geometry that gives the required distribution of melt in the extrudate. At the same time, the pressure required at the entry to the die must not exceed that readily available from the extruder. Also, the die body must be robust enough for the deformations produced by the internal pressures to cause insignificant changes in the flow-channel dimensions. Finally, it is necessary to design a temperature control system for the die that allows the flow-channel boundaries to be maintained at or near certain set temperatures, which are often close to the temperature of the incoming melt. For present purposes, attention is confined to achieving the required flow distribution, and determining at least the over-all pressure difference.

In section 5.1, there was a discussion of some simple one-dimensional die flows in which the flow distribution was known; the main emphasis was on calculating pressure differences. With the two-dimensional narrow channel flows considered in section 5.2, however, determination of the flow distribution required a more sophisticated type of analysis. In the following subsections, two typical die design problems are examined, involving the use of the two forms of analysis.

5.3.1 Design of a Flat-film Die

The principles of the coat-hanger type of flat-film die were discussed in section 2.1.4. Analytical results obtained in section 5.1 can now be used to guide the choice of the main dimensions of such a die. Figure 5.10a shows the cross-section through the flow channel in the mid plane of the extruded film, while figure 5.10b shows a typical section normal to the die lips, on the line marked AA in figure 5.10a. The design problem is to choose a set of flow channel dimensions that give a uniform distribution of film thickness at the die exit, with a minimum amount of die lip adjustment. This is achieved by ensuring that the pressure drop along any flow path is independent of that path. To this end, it is reasonable to assume that flow along the circular manifold channel is parallel to its axis, while the direction of flow in the tapering flat channel between the die lips is parallel to the direction of film extrusion. Hence, the pressure drop from the point B in the melt inlet to point C directly opposite the inlet should be the same as that from B along the manifold to E, and then to the die exit at F.

Suppose that the two arms of the manifold are straight, and inclined at angles β as shown. Also, the diameter of the manifold, D, is taken to be constant, as is the distance, H_0, between the die lips at the exit. If the angle of taper of the die lips is 2α, as shown in figure 5.10b, the distance between them, H, at the manifold is the following function of the lateral distance, x, from the axis of the melt inlet channel

$$H = H_0 + 2L \tan \alpha \tag{5.61}$$

where

$$L = L_1 - x \tan \beta \tag{5.62}$$

is the length of the die lips in the direction of flow, and L_1 is its maximum value at the centre of the die. Now, if Q is the total volumetric flow rate of melt through the die, the required constant rate per unit width of the die lips is $q = Q/W$, where W is their over-all width. Also, the flow rate along the manifold reduces linearly with

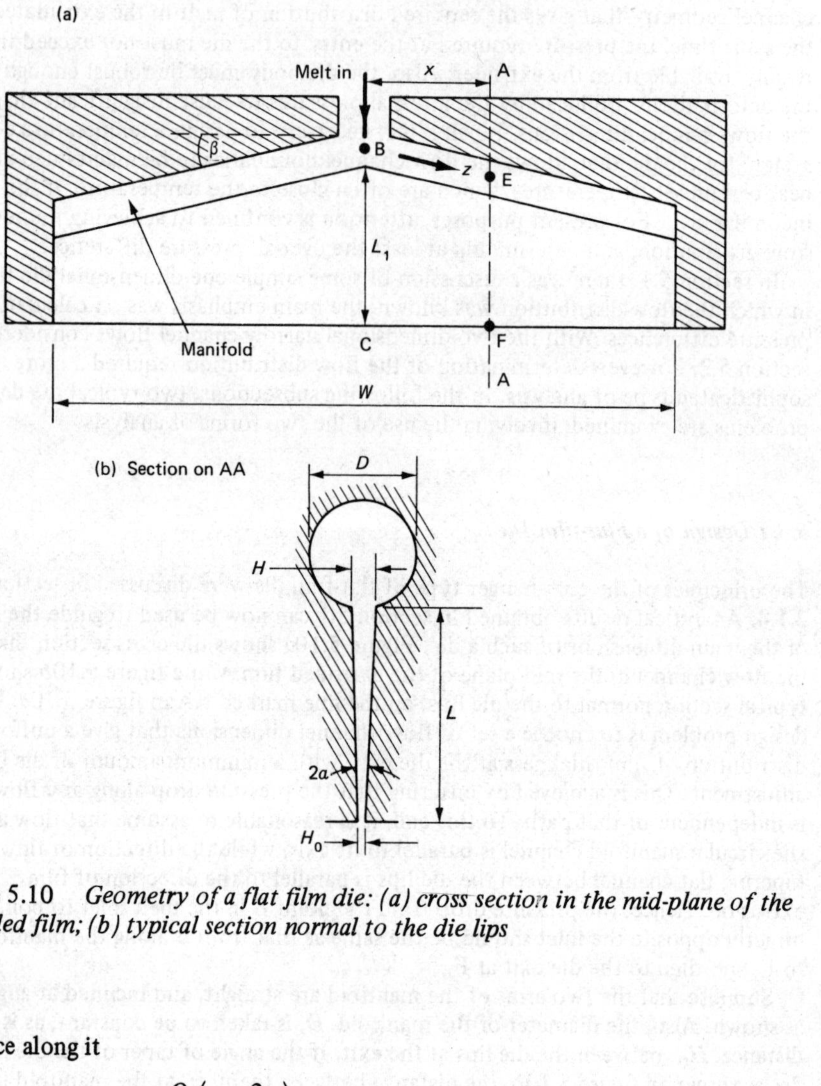

Figure 5.10 *Geometry of a flat film die: (a) cross section in the mid-plane of the extruded film; (b) typical section normal to the die lips*

distance along it

$$Q' = \frac{Q}{2}\left(1 - \frac{2x}{W}\right) \tag{5.63}$$

The pressure difference between points B and C can be established with the aid of equation 5.29 as

$$p_B - p_C = -\frac{P_{x1}L_1H_1}{2n(H_1 - H_0)}\left[\left(\frac{H_0}{H_1}\right)^{-2n} - 1\right] \tag{5.64}$$

where H_1 is the value of H given by equation 5.61 when $x = 0$, $L = L_1$, and P_{x1} is the pressure gradient in a flat channel of depth H_1 carrying a flow rate per unit width width of q, according to equation 5.25. Similarly, the pressure difference between points E and F is

$$p_E - p_F = -\frac{P_x LH}{2n(H - H_0)}\left[\left(\frac{H_0}{H}\right)^{-2n} - 1\right]$$ (5.65)

where P_x is the pressure gradient in a channel of depth H carrying the same flow rate. Now, the calculation for the pressure difference between B and E must take into account the reducing flow rate along the manifold. Introducing coordinate z along the axis of the manifold arm under consideration, the initial pressure gradient, P_{z1}, can be found from equation 5.11 for a volumetric flow rate $Q/2$ along a circular tube of diameter D. Hence, according to equation 5.9, the pressure gradient at a position where the flow rate is given by equation 5.63 is

$$P_z = P_{z1}\left(1 - \frac{2x}{W}\right)^n$$ (5.66)

and the pressure difference between points B and E is

$$p_B - p_E = -\int_B^E P_z\,dz = \frac{WP_{z1}}{2(n + 1)\cos\beta}\left[\left(1 - \frac{2x}{W}\right)^{n+1} - 1\right]$$ (5.6$^\cdot$

As the film leaves the die under atmospheric conditions, $p_C = p_F = 0$, and the value of p_B obtained from equation 5.64 should be equal to that obtained by adding together equations 5.65 and 5.67 to eliminate p_E. Hence, given the power-law index of the polymer and dimensions such as H_0, W, L_1, D and β, it is possible to determine the required functional form of $\alpha(x)$ to give uniform film thickness. Varying the taper angle across the width of the die is not an ideal arrangement, both because the melt leaving the die should all have suffered the same recent deformation history, and because such a die is not easy to manufacture. An alternative approach is therefore to fix α, and choose β to give the best — though by no means perfect — thickness distribution. If the manifold arms are allowed to be curved, the die-length distribution $L(x)$ required to satisfy the pressure balance equations can be calculated. Yet another possibility is to allow the die gap, H, to vary nonlinearly, for example, by means of more than one taper of different angles. Clearly, there are a number of possible solutions to the design problem, any one of which can be established by the application of an essentially straightforward set of algebraic flow equations. Probably the most practical approach is to select a form of geometry that is easy and cheap to manufacture, even if it does not produce a perfect thickness distribution according to the analysis. Provision for fine adjustments will have to be made in any case to cater for minor material property and temperature variations, as well as die body deformations under pressure. The magnitudes of such deformations can, in principle, be estimated from the melt pressure distribution predicted by the flow analysis.

5.3.2 Improved Design of Cable Covering Crosshead Deflector

A method of analysing polymer melt flow in the narrow channel of a cable covering crosshead deflector was described in section 5.2.2. Despite the contouring of the boundaries of the channel, the flow distribution produced by the particular deflector studied was far from perfect, and was only made acceptable by the considerable throttling effect of the tapered point attached to the deflector. Clearly, there is scope for improving the deflector design.

In principle, the method of flow analysis can be inverted so that, given a desired pattern of streamlines, the corresponding channel-depth distribution can be computed. In practice, however, such a design procedure leads to a flow channel geometry of complex shape, which is difficult and therefore expensive to manufacture. A more practical but more empirical approach is to make geometrically simple modifications to the existing deflector design, the effects of which on the flow distribution can be examined with the aid of the flow analysis.

Figure 5.9 shows that, using the deflector without the point, considerably more melt flows in the middle of each channel — and so across boundary CD in figure 5.8 — than at the sides. Therefore, the distribution should be made more uniform by directing melt outwards towards these sides. Bearing in mind that the channel reduces linearly in depth in the axial direction, one possible way to do this is to introduce a triangular wedge shaped region, as shown in figure 5.11, where the channel depth is constant and equal to the depth at boundary CD. Thus, the difference in depth between the wedge region and the main channel is greatest at the apex point marked I, and reduces to zero along the line CGHD. In manufacturing terms, this simply means that there is no axial taper in the channel depth in the wedge region GHI.

For the particular deflector and polymer considered in the earlier example, the best design of this type may be found by varying the wedge angles, and is illustrated

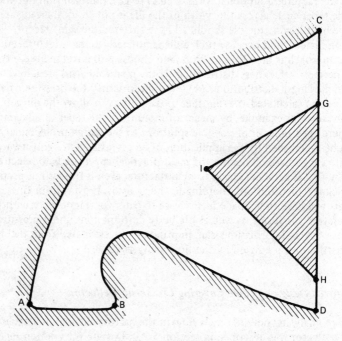

Figure 5.11   *Deflector flow channel with wedge to improve flow distribution*

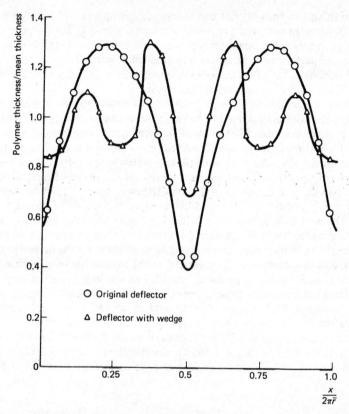

Figure 5.12 *Computed final thickness distributions for deflector alone, both with and without the wedge*

in figure 5.11. The considerable predicted improvement in thickness distribution produced by the deflector alone is shown in figure 5.12. Even with relatively little throttling produced by the point, the polymer layer thickness distribution on the finished cable should be acceptable. It should be mentioned, however, that one of the most severe difficulties encountered in practice is to ensure correct alignment of the point with the deflector, to which it is often attached by a threaded joint. Only a minute misalignment is necessary to severely distort the final thickness distribution, an effect which can, however, be studied using the present method of flow analysis.

5.4 Calendering

As indicated in the general description of calendering in section 2.4, the process involves forcing molten material through the narrow gaps, or nips, between rotating cylindrical rolls in order to produce a sheet. At the same time, there is a certain amount of mixing of the polymer and also heating, which occurs both by conduction from the rolls and as a result of internal shearing. The purpose of attempting to analyse the calendering process is to establish relationships between the design and performance parameters such as roll diameters, speeds, size of nip and sheet thickness. Melt flow between calender rolls is, in many respects, similar to the other flows already considered in this chapter.

The heart of calendering is the melt flow in the region of the nip shown in figure 5.13. The rolls are assumed to be of the same radius, R, but may have different peripheral speeds, U_1 and U_2. Cartesian coordinates x and y, with an origin at the centre of the nip, are chosen to be parallel and normal to the flow direction, respectively. The minimum distance between the rolls is H_0 at the nip itself, increasing to H_1 at the position where the sheet separates from the rolls, a distance L_1 in the x direction from the nip. Although figure 5.13 shows separation occurring simultaneously from the two rolls, in practice the sheet may adhere to one of them. The melt flow before entry to the nip region may take the form of a recirculating rolling bank. For present purposes, it is assumed that the rolls are rigid enough for the flow-boundary geometry to be unaffected by the pressures generated in the nip, also that there is no flow parallel to the axes of the rolls, normal to the plane shown. As $H_0 \ll R$, the geometric conditions for the lubrication approximation to be applicable are satisfied close to the nip, but not in the earlier region of the rolling bank where the flow is two dimensional. If isothermal conditions are assumed, attention can be confined to solving for the flow velocities.

Apart from the change in velocity boundary conditions, melt flow between calender rolls is similar to that in a flat slit die described in section 5.1.2. Equilibrium equation 4.73 can be applied at a typical flow cross-section a distance

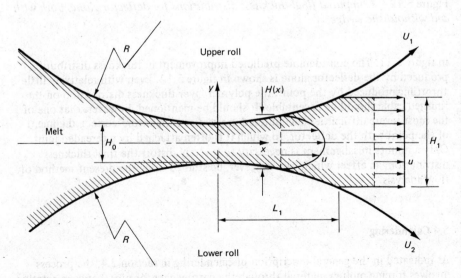

Figure 5.13 *Melt flow in the nip between calender rolls*

x from the nip, where the gap between the rolls is $H(x)$

$$\frac{d\tau_{xy}}{dy} = \frac{dp}{dx} \quad \text{with} \quad \begin{array}{l} u = U_1 \text{ at } y = +H/2 \\ u = U_2 \text{ at } y = -H/2 \end{array} \tag{5.68}$$

The following simplified form of $H(x)$ is consistent with the lubrication approximation

$$H = H_0 + \frac{x^2}{R} \tag{5.69}$$

Before attempting to solve equation 5.68 for the case of a power-law constitutive equation, it is worth considering a newtonian flow analysis. Although this does not give results that are quantitatively reliable, it does demonstrate the important features of the flow without involving too much analytical complexity.

5.4.1 Newtonian Analysis of Calendering

If the melt is assumed to have constant viscosity, μ, equation 5.68 becomes

$$\mu \frac{d^2 u}{dy^2} = \frac{dp}{dx} \equiv P_x \tag{5.70}$$

which may be integrated subject to the boundary conditions to give the velocity profile

$$u = \frac{P_x}{2\mu} \left[y^2 - \left(\frac{H}{2} \right)^2 \right] + \frac{y}{H} (U_1 - U_2) + \frac{U_1 + U_2}{2} \tag{5.71}$$

from which the volumetric flow rate per unit width may be obtained as

$$Q = \int_{-H/2}^{+H/2} u \, dy = \frac{H(U_1 + U_2)}{2} - \frac{P_x H^3}{12\mu} \tag{5.72}$$

Note that this result is a simple superposition of a drag flow rate term due to the mean roll velocity, $(U_1 + U_2)/2$, on the flow rate due to the pressure gradient previously derived in equation 5.23 for the stationary boundary case. In order to satisfy continuity requirements for steady flow, Q must be independent of x, and equation 5.72 may be rearranged to define the pressure gradient as a function of x

$$\frac{dp}{dx} = \frac{6\mu(U_1 + U_2)}{H^2} - \frac{12\mu Q}{H^3} \tag{5.73}$$

An appropriate choice of pressure boundary conditions, which allows both the pressure profile and the flow rate to be determined, is

$$p = \frac{dp}{dx} = 0 \text{ at } x = L_1 \tag{5.74}$$

In other words, the pressure in the melt must be ambient at the position where separation from one or both rolls occurs. The zero pressure gradient condition at the same position is an assumption commonly used in hydrodynamic lubrication analysis. It implies that the pressure gradient is continuous at the separation, which means that there is not an abrupt change from a curved velocity profile in the roll gap to a flat one in the sheet produced. Using this condition, equation 5.73 yields the following expression for the flow rate

$$Q = H_1 \frac{U_1 + U_2}{2} \tag{5.75}$$

and equation 5.73 becomes

$$\frac{dp}{dx} = 6\mu(U_1 + U_2)\left(\frac{1}{H^2} - \frac{H_1}{H^3}\right) \tag{5.76}$$

Given the zero pressure boundary condition and the dependence of H on x defined in equation 5.69, this can be integrated to give

$$p = 6\mu(U_1 + U_2)\frac{\sqrt{(RH_0)}}{H_0{}^2}[\phi(\theta) - \phi(\theta_1)] \tag{5.77}$$

where

$$\phi(\theta) = \frac{1}{4}(2\theta + \sin 2\theta) - \frac{H_1}{32H_0}(12\theta + 8\sin 2\theta + \sin 4\theta)$$

$$\theta = \tan^{-1}\left[\frac{x}{\sqrt{(RH_0)}}\right], \theta_1 = \tan^{-1}\left[\frac{L_1}{\sqrt{(RH_0)}}\right]$$

The form of this pressure profile is shown in figure 5.14, together with typical velocity profiles associated with different regions of the flow. Note that the maximum pressure is achieved at a distance L_1 before the nip, where the roll gap is H_1 and the pressure gradient zero. The maximum pressure gradient occurs at the nip. The position at which pressure generation starts, shown at a distance L_0 before the nip, can be regarded as the effective entry to the nip region.

Having determined the pressure profile, the total force acting in the y direction on each roll can be obtained as

$$F = \int_{-L_0}^{+L_1} p \, dx \tag{5.78}$$

per unit length of roll. Similarly, the power required to turn, say, the upper roll can be calculated as the rate of working at the interface between the roll and the melt. For unit length of roll this is

$$E_1 = \int_{-L_0}^{+L_1} U_1(\tau_{xy})_{y=H/2} \, dx = \int_{-L_0}^{+L_1} U_1\left(\frac{P_x H}{2} + \mu\frac{U_1 - U_2}{H}\right) dx \tag{5.79}$$

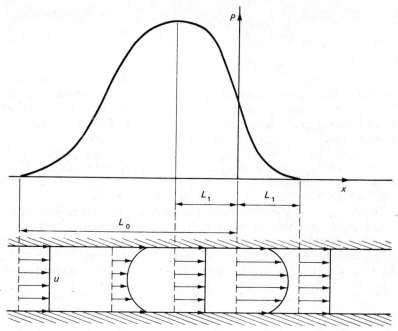

Figure 5.14   *Pressure profile in a calender nip*

5.4.2 Non-newtonian Analysis of Calendering

If the newtonian flow assumption is rejected, and the melt viscosity is assumed to be the power-law function of the local rate of shear defined by equation 5.18, equation 5.68, may be integrated to give

$$\mu_0 \left| \frac{1}{\gamma_0} \frac{du}{dy} \right|^{n-1} \frac{du}{dy} = P_x(y - y_0) \tag{5.80}$$

where $y = y_0$ is the position of the stress neutral point at the particular flow cross section. As discussed in section 5.1.3, further analytical progress is difficult when y_0 is at some unknown position within the flow.

For the particular case of equal roll speeds, $U_1 = U_2 = U$, say, the flow is symmetrical about $y = 0$, and therefore $y_0 = 0$. Hence, for the upper half of the flow, $y \geqslant 0$, equation 5.80 can be rearranged as

$$\frac{du}{dy} = \text{sgn}\,(P_x) \left(\frac{\gamma_0^{\,n-1}}{\mu_0} |P_x| \right)^{1/n} y^{1/n} = C_5 y^{1/n} \tag{5.81}$$

from which

$$u = \frac{C_5}{\left(\frac{1}{n} + 1 \right)} \left[y^{1/n+1} - \left(\frac{H}{2} \right)^{1/n+1} \right] + U \tag{5.82}$$

and

$$Q = 2 \int_0^{H/2} u \, dy = UH - \frac{2C_5}{\left(\dfrac{1}{n}+2\right)} \left(\frac{H}{2}\right)^{1/n+2} \tag{5.83}$$

Although an expression for the local pressure gradient can be obtained from this result, integration to find the pressure profile can only be carried out numerically. The general form of this profile is again as shown in figure 5.14. Roll forces and power consumptions can also be found by numerical integration.

For the more general case of $U_1 \neq U_2$, even local velocity profiles must be obtained numerically. Dimensionless parameters of the types first discussed in section 4.5 for characterising melt flows can be defined for flow between calender rolls. The nip gap H_0 and length L_1 are appropriate flow channel dimensions. While the mean velocity $(U_1 + U_2)/2$ should be used to find characteristic Reynolds, Peclet and Graetz numbers, the mean shear rate is in general best defined as $(U_1 - U_2)/H_0$, that is, in terms of the difference in peripheral speeds of the rolls. Therefore, while the Peclet and Graetz numbers may be high, the Griffith number is likely to be relatively low, and the isothermal flow assumption reasonably valid.

5.4.3 Mixing in a Calender Nip

As one of the functions of the calendering process is to mix the melt by shearing before forming a sheet, it is relevant to apply the distributive mixing analysis developed in section 4.7 to the flow in the region of the nip. Using equation 4.84, the bulk mean mixing per unit length in the direction of flow is

$$\frac{d\overline{M}}{dx} = \frac{1}{Q} \int_{-H/2}^{+H/2} (4I_2)^{1/2} \, dy \tag{5.84}$$

where for the simple shear flow involved

$$I_2 = \frac{1}{4} \left(\frac{du}{dy}\right)^2 \tag{5.85}$$

For present purposes, it is sufficient to consider mixing in newtonian flow, for which the required velocity gradient may be obtained from equation 5.71 as

$$\frac{du}{dy} = \frac{yP_x}{\mu} + \frac{U_1 - U_2}{H} \tag{5.86}$$

Using equations 5.75 and 5.76 for the flow rate and pressure gradient, respectively, the mixing gradient defined by equation 5.84 becomes

$$\frac{d\overline{M}}{dx} = \int_{-H/2}^{+H/2} \left| 12\left(\frac{1}{H^2} - \frac{H_1}{H^3}\right) \frac{y}{H_1} + \frac{2(U_1 - U_2)}{H_1 H(U_1 + U_2)} \right| dy \tag{5.87}$$

Note that this result is independent of melt viscosity. The mixing imparted is inversely proportional to the size of the gap between the rolls. For example, consider only the mixing due to the difference in the roll speeds

$$\frac{d\bar{M}}{dx} = \frac{2 \mid U_1 - U_2 \mid}{H_1(U_1 + U_2)}$$ (5.88)

Mixing is improved by increasing the speed difference. Even better mixing could be obtained by reversing the direction of rotation of one roll, perfect mixing (but no flow rate) being achieved when $U_2 = -U_1$.

5.5 Melt Flow in an Intensely Sheared Thin Film

In most processing operations, polymer is melted by a combination of heat conducted from a hot metal surface, and intense shear in the thin film of melt formed between the surface and a compacted bed of solid granules or powder. For example, such a mechanism occurs in screw extruders, and is discussed in chapter 6. Therefore, in preparation for the modelling of melting in extruders, this section is concerned firstly with the analysis of nonisothermal drag flow in a thin film, and secondly with the effect of an influx of freshly melted material from one of the flow boundaries. The justification for assuming drag flow, thus ignoring the effects of pressure gradients, is that because the film is thin, the magnitude of a pressure gradient large enough to influence the velocity profile significantly would require pressure differences too large to be sustained by the solid bed of polymer. An important advantage of this simplification is that it makes possible a considerable amount of analytical progress towards a solution, even when the flow must be treated as nonisothermal.

Figure 5.15 illustrates the problem to be solved, in which the lower boundary is treated as stationary and at a known temperature, T_m, typically the melting point of the polymer. The upper boundary is maintained at temperature T_b, and moves at relative velocity $\bar{U}$ in the x direction parallel to the boundaries. This notation for velocity is chosen deliberately, to conform with that used in section 4.5. Coordinate y is normal to the boundaries, the local distance between which is H.

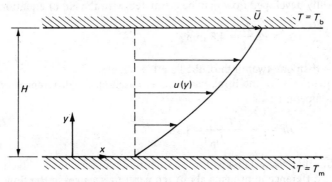

Figure 5.15   *Melt flow in a thin film*

In order to establish the type of flow regime involved in terms of characteristic dimensionless parameters, consider a particular case in which

$$H = 0.4 \text{ mm}, \overline{U} = 100 \text{ mm/s}, T_m = 150\,°C, T_b = 170\,°C$$

and the polymer concerned has the following properties at its melting point

$$\mu_0 = 50.0 \text{ kN s/m}^2 \text{ at } \gamma_0 = 1 \text{ s}^{-1}, n = 0.5, b = 0.015\,°C^{-1}$$
$$\rho = 750 \text{ kg/m}^3, C_p = 2000 \text{ J/kg }°C, k = 0.50 \text{ W/m }°C$$

Hence, the mean shear rate, viscosity and shear stress at the same temperature are

$$\overline{\gamma} = \frac{\overline{U}}{H} = 250 \text{ s}^{-1}, \overline{\mu} = \mu_0 \left| \frac{\overline{\gamma}}{\gamma_0} \right|^{n-1} = 3.16 \text{ kN s/m}^2$$
$$\overline{\tau} = \overline{\mu}\overline{\gamma} = 791 \text{ kN/m}^2$$

and the Reynolds, Peclet and Griffith numbers are

$$Re = \frac{\rho\overline{U}H}{\overline{\mu}} = 9.49 \times 10^{-6}$$

$$Pe = \frac{\rho C_p \overline{U}H}{k} = 120$$

$$G = \frac{b\overline{\tau}\overline{\gamma}H^2}{k} = 0.949$$

While the small size of Re ensures that inertia effects are negligible, the Griffith number is such that the flow cannot be assumed to be isothermal. Although the size of the Peclet number suggests that thermal convection may be important, the Graetz number based on a film length of $L = 100$ mm in the direction of flow is

$$Gz = Pe\,\frac{H}{L} = 0.48$$

Heat conduction to and from the boundaries in thermally fully developed flow is the dominant mode of heat transfer. Put another way, the length of flow necessary to achieve fully developed flow may be estimated with the aid of equation 4.69 as

$$L' = Pe\,\frac{H}{10} = 4.8 \text{ mm}$$

which is less than one-twentieth of the length available.

Using equation 4.68, the Brinkman number based on the difference between the boundary temperatures is

$$Br = \frac{\overline{\tau}\overline{\gamma}H^2}{k(T_b - T_m)} = 3.16$$

which implies that viscous dissipation is rather more important than the boundary-temperature difference in bringing about temperature changes in the flow. In order for a pressure gradient in the direction of flow to have a significant effect on velocities, the dimensionless pressure gradient defined in equation 4.61 would need to be of order unity. That is

$$\frac{\partial p}{\partial x} \approx \frac{\bar{\tau}}{H} = 1980 \text{ MN/m}^3$$

implying a pressure difference of the order of 200 MN/m^2 over the 100 mm length of flow. Such a difference is much larger than any observed in practice, and too large to be sustained by the compacted solid polymer forming the lower flow boundary. The drag flow assumption is therefore justified, as is the application of the lubrication approximation to both velocities and temperatures.

5.5.1 Nonisothermal Drag Flow of a Power-law Temperature Sensitive Melt

Proceeding with the analysis of the drag flow illustrated in figure 5.15, if the pressure gradient in the x direction is negligible, equilibrium equation 4.73 yields

$$\tau_{xy} = \text{const} = \tau, \text{ say} \tag{5.89}$$

Energy equation 4.76 may be used to determine the temperature profile

$$k \frac{d^2 T}{dy^2} = -\tau \frac{du}{dy} \tag{5.90}$$

Power-law constitutive equation 4.38, together with the temperature dependence of viscosity displayed in equation 3.19, gives the following relationship between τ, the velocity gradient and temperature

$$\tau = \mu_0 \left| \frac{1}{\gamma_0} \frac{du}{dy} \right|^{n-1} \frac{du}{dy} \exp\left[-b(T - T_0)\right] \tag{5.91}$$

Now, if the boundary velocity, $\bar{U}$, is positive, then so are τ and the velocity gradient throughout the flow. It is convenient to introduce the following dimensionless coordinate, velocity, temperature and shear stress

$$Y = \frac{y}{H}, U = \frac{u}{\bar{U}}, T^* = b(T - T_m), S = \frac{\tau}{\bar{\tau}} \tag{5.92}$$

where the mean stress, $\bar{\tau}$, is defined at the mean shear rate, $\bar{\gamma} = \bar{U}/H$, and the temperature, T_m, of the lower boundary

$$\bar{\tau} = \mu_0 \left(\frac{1}{\gamma_0} \frac{\bar{U}}{H} \right)^{n-1} \frac{\bar{U}}{H} \exp\left[-b(T_m - T_0)\right] \tag{5.93}$$

Therefore, equations 5.90 and 5.91 become

$$\frac{d^2 T^*}{dY^2} = -GS \frac{dU}{dY} \tag{5.94}$$

$$S = \left(\frac{dU}{dY} \right)^n \exp\left(-T^*\right) \tag{5.95}$$

where G is the Griffith number, as previously defined. Rearranging equation 5.95 for the dimensionless velocity gradient

$$\frac{dU}{dY} = S^{1/n} \exp\left(\frac{T^*}{n}\right) \tag{5.96}$$

which may be substituted into equation 5.94 to give

$$\left(\frac{dT^*}{dY}\right) \frac{d}{dT^*}\left(\frac{dT^*}{dY}\right) = -GS^{1/n+1} \exp\left(\frac{T^*}{n}\right) \tag{5.97}$$

Hence

$$\left(\frac{dT^*}{dY}\right)^2 = 2nGS^{1/n+1}\left[\exp\left(\frac{T_0^*}{n}\right) - \exp\left(\frac{T^*}{n}\right)\right] \tag{5.98}$$

where T_0^* is defined such that $dT^*/dY = 0$ at $T^* = T_0^*$. In order to integrate again to find T^*, the following substitution

$$\cosh^2\phi = \exp\left(\frac{T_0^* - T^*}{n}\right) \tag{5.99}$$

can be used to give

$$\frac{d\phi}{dY} = BS^{(1+n)/2n}A \tag{5.100}$$

where

$$A = \exp\left(\frac{T_0^*}{2n}\right), \qquad B = \left(\frac{G}{2n}\right)^{1/2} \tag{5.101}$$

As the right-hand side of equation 5.100 is independent of Y

$$\phi = BS^{(1+n)/2n}AY + C \tag{5.102}$$

where C is an integration constant. Hence, substituting into equation 5.99

$$A^2 \exp\left(-\frac{T^*}{n}\right) = \cosh^2(BS^{(1+n)/2n}AY + C) \tag{5.103}$$

which defines the dimensionless temperature profile as a function of Y.

Two relationships between the unknowns A, C and S are provided by the thermal boundary conditions

$$T^* = 0 \text{ at } Y = 0; A = \cosh C \tag{5.104}$$

$$T^* = b(T_b - T_m) \text{ at } Y = 1; E \cosh^2 C = \cosh^2 F \tag{5.105}$$

where

$$E = \exp\left(-b\frac{T_b - T_m}{n}\right) \tag{5.106}$$

$$F = BS^{(1+n)/2n}A + C = BS^{(1+n)/2n}\cosh C + C \tag{5.107}$$

Equation 5.105 represents a single relationship between the unknowns S and C.

Returning to the velocity analysis, equation 5.103 can be used in equation 5.96 to give

$$\frac{dU}{dY} = S^{1/n}A^2 \, \text{sech}^2 \, (BS^{(1+n)/2n}AY + C) \tag{5.108}$$

Integration and application of the velocity boundary condition $U = 0$ at $Y = 0$ yields

$$U = \frac{AS^{(1-n)/2n}}{B} \, [\tanh \, (BS^{(1+n)/2n}AY + C) - \tanh C] \tag{5.109}$$

The second velocity boundary condition, $U = 1$ at $Y = 1$, provides another relationship between the unknowns S and C

$$1 = \frac{S^{(1-n)/2n} \cosh C}{B} (\tanh F - \tanh C) \tag{5.110}$$

Although non-linear equations 5.105 and 5.110 must now be solved numerically for S and C, they are at least algebraic rather than differential equations. This substantially reduces the amount of computation involved, which is an important consideration when the present method of analysis is used repeatedly in the modelling of the melting process in screw extruders.

Having found S and C, the former yielding the shear stress in the melt film, it is a straightforward matter to determine, for example, the temperature gradient at the lower boundary, which is important for calculating the melting rate there. Differentiation of equation 5.103 with respect to Y gives

$$\left(\frac{dT^*}{dY} \right)_{Y=0} = -2nBS^{(1+n)/2n} \sinh C \tag{5.111}$$

Also, the dimensionless volumetric flow rate is

$$\pi_Q = \int_0^1 U \, dY = \frac{AS^{(1-n)/2n}}{B} \left[\frac{1}{ABS^{(1+n)/2n}} \ln \left(\frac{\cosh F}{\cosh C} \right) - \tanh C \right] \tag{5.112}$$

where $\pi_Q = Q/\bar{U}H$, Q being the actual volumetric flow rate.

As an illustration of the use of the above results, consider the numerical example employed in the introduction to thin film flow analysis. Only three parameters are necessary to define the problem expressed in dimensionless form; these are

$$n = 0.5, \qquad G = 0.949, \qquad E = \exp \left(-b \frac{T_b - T_m}{n} \right) = 0.549$$

Numerical solution of simultaneous equations 5.105 and 5.110 gives

$$S = 0.800, \qquad C = -0.825$$

This result for S implies that the actual shear stress in the film is only 80 per cent of the characteristic shear stress defined at the mean shear rate and the temperature of the lower flow boundary

$$\tau = S\bar{\tau} = 0.800 \times 791 = 633 \text{ kN/m}^2$$

Also, equations 5.111 and 5.112 yield

$$\left(\frac{dT^*}{dY}\right)_{Y=0} = 0.643, \ \pi_Q = 0.452$$

For comparison, in isothermal drag flow when viscous dissipation makes no contribution to the temperature profile, both the velocity and temperature profiles would be linear and the corresponding values of the parameters would be

$$\left(\frac{dT^*}{dY}\right)_{Y=0} = b(T_b - T_m) = 0.3, \ \pi_Q = 0.5$$

Clearly, the effect of viscous dissipation is to more than double the temperature gradient, while the coupling of the temperature and velocity profiles via the temperature sensitive viscosity reduces the flow rate by some 10 per cent. Figure 5.16 shows the velocity and temperature profiles, derived from equations 5.109 and 5.103, respectively, the dotted lines indicating the simple linear profiles. The maximum temperature in the flow is slightly higher than the upper boundary temperature. Viscous dissipation not only more than doubles the rate of heat conduction to the lower boundary, where melting is occurring, but also causes a small amount of heat to be conducted to the hotter upper boundary. This effect was anticipated by the earlier calculation of a characteristic Brinkman number for the flow.

5.5.2 Melting into a Thin Film

The main application for the above analysis of drag flow in a thin melt film is to melting in screw extruders. Such melting involves an influx of freshly melted material, which progressively increases the flow rate of melt in the film. This effect can be accommodated in a numerical treatment of nonisothermal film development by the method described in section 6.3.3. For present purposes, however, it is

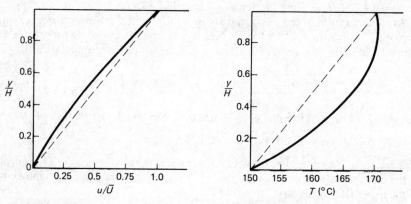

Figure 5.16   Velocity and temperature profiles in thin film example

instructive to consider the simplified analysis of isothermal film flow. In isothermal flow, the velocity profile is linear, and $du/dy = \bar{\gamma}$, $\tau = \bar{\tau}$ throughout the depth of the film. Equation 5.90 becomes

$$\frac{d^2 T}{dy^2} = -\frac{\bar{\tau}\bar{\gamma}}{k} = -\frac{G}{bH^2} \tag{5.113}$$

In strictly isothermal flow, G is zero, but the essential feature of such flow has already been imposed by uncoupling the velocity and temperature profiles. Integrating equation 5.113 with the present boundary conditions

$$\left(\frac{dT}{dy}\right)_{y=0} = \frac{G}{2bH} + \frac{T_b - T_m}{H} \tag{5.114}$$

Note, incidentally, that the dimensionless form of this result is

$$\left(\frac{dT^*}{dY}\right)_{Y=0} = bH\left(\frac{dT}{dy}\right)_{y=0} = \frac{G}{2} + b(T_b - T_m) \tag{5.115}$$

and in the above numerical example

$$\left(\frac{dT^*}{dY}\right)_{Y=0} = \frac{0.949}{2} + 0.015 \times 20 = 0.774$$

which compares with 0.643 derived from the nonisothermal analysis.

Bearing in mind the melting behaviour of polymers discussed in section 3.1, the material is assumed to have a latent heat of fusion, λ, at its melting point, T_m. Therefore, the volumetric rate of influx of melted material, q per unit area of the melt and solid interface, is given by

$$\rho \lambda q = k\left(\frac{dT}{dy}\right)_{y=0} = \frac{k}{H}\left(\frac{G}{2b} + T_b - T_m\right) \tag{5.116}$$

This result assumes that the temperature of all the solid material is at the melting point, and neglects the heat required to raise the temperature of the newly formed melt to the mean temperature of the film. Corrections for these effects are readily made by modifying λ, as described in section 6.3.3. The rate of increase in the direction of flow of the volume of material flowing is

$$\frac{dQ}{dx} = q = \frac{k}{\rho \lambda H}\left(\frac{G}{2b} + T_b - T_m\right) \tag{5.117}$$

Now, by virtue of the linear velocity profile, $Q = \bar{U}H/2$. Since $\bar{U}$ does not vary with x, the film thickness H must increase to accommodate the additional material. Equation 5.117 becomes

$$\frac{dH}{dx} = \frac{2k}{\rho \lambda H \bar{U}}\left(\frac{G}{2b} + T_b - T_m\right) \tag{5.118}$$

which, if G is assumed to remain constant, may be integrated to give

$$H = \left[\frac{4k}{\rho\lambda\overline{U}} \left(\frac{G}{2b} + T_b - T_m \right) x + H_0{}^2 \right]^{1/2} \qquad (5.119)$$

where $H = H_0$ at $x = 0$. In other words, the film thickness grows approximately in proportion to the square root of the distance in the direction of flow.

For example, in the previously considered numerical problem, suppose that $\lambda = 100$ kJ/kg, and that the film grows from its initial thickness of 0.4 mm over its length of 100 mm. The final thickness may be estimated as

$$H = \left[\frac{4 \times 0.50}{750 \times 100 \times 10^3 \times 100 \times 10^{-3}} \left(\frac{0.949}{2 \times 0.015} + 20 \right) 0.1 \right.$$
$$\left. + (0.4 \times 10^{-3})^2 \right]^{1/2} \text{m}$$

$$= 1.24 \text{ mm}$$

although the value of G would have changed significantly over the length of the film. A more accurate analysis would take into account both this effect and the nonisothermal coupling of velocities and temperatures.

6

Screw Extrusion

In this chapter, attention is confined to screw extrusion, arguably the most important polymer process. This importance is reflected in the amount of the effort that has been devoted to trying to understand and analyse melt and solid flow in extruders, particularly single-screw extruders. Having established in chapter 4 the continuum mechanics equations governing flow of polymer melts, and illustrated their application to relatively simple melt flow processes in chapter 5, the analysis of flow in screw channels can now be undertaken. Except in the case of melt-fed homogenising extruders, melt flow in the absence of solid polymer only occurs towards the delivery end of the machine. Conveying of solid material and melting under the combined effects of conducted and generated heat are important stages of the extrusion process. Early attempts to analyse extruder performance were, however, concentrated on melt flow. It is for this reason, and also because the analysis of melting involves treatments of melt flow, that the latter is considered first here.

Concentrating on single-screw extruders, such as the typical one shown in figure 2.1, figure 6.1 shows an enlarged part of the screw and barrel, and serves to

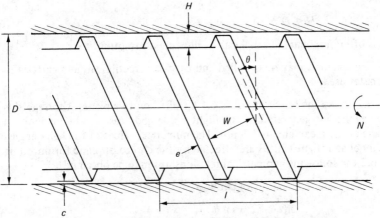

Figure 6.1   *Geometry of an extruder screw (two channels in parallel)*

define the geometry. The internal barrel diameter is D, and the channel depth, H, is the radial gap between the barrel and screw root, while c is the radial clearance between screw flight and barrel. The channel and flight widths, W and e, respectively are measured normal to the flights along the cylindrical surface generated by the flight tips. The helix angle, θ, is also measured at the flight tips, and l is the flight lead. The number of channels in parallel, or the number of separate flights or 'starts' on the screw, is m. Since $m = 2$ in figure 6.1, the lead is twice the flight pitch, which defines the axial distance between successive flights. The screw rotates in the direction shown, at N revolutions per unit time.

6.1 Melt Flow in Screw Extruders

In the development of analyses for melt flow in the channels of single-screw extruders, a number of assumptions are normally made, some of which are reasonable, others much less so. In the following derivation, these assumptions are numbered consecutively as they occur. For example, there are three geometric assumptions.

(1) The sides of the flights are radial to the screw axis.
(2) The depth of the screw channel is constant across its width.
(3) Chamfers or fillets on the flights may be neglected.

All three are valid for most screws.

A coordinate system fixed relative to the screw is selected, since it is simple to use. A further simplification is to treat the barrel as rotating about a stationary screw, which is a valid procedure for the following reasons.

(4) Body forces, such as those due to gravity, are negligible in comparison with viscous forces.
(5) Centrifugal inertia forces are likewise negligible.

The most natural coordinate system to use is a helical one. In the melt flow regions of most single-screw extruders, however, the screw channels are relatively shallow, such that

$$H \ll D \tag{6.1}$$

and it is often reasonable to make the following assumption.

(6) The channels may be unrolled and treated as rectilinear, and cartesian coordinates used.

Figure 6.2 shows the screw of figure 6.1 unrolled on to a plane. Although this plane is taken as the one generated by the flight tips, a good case can be made for choosing the plane at the mean channel depth. The numerical values of the helix angle and the channel and flight widths are affected by the choice of plane for unrolling. From figure 6.2 the following geometrical relationships may be obtained

$$l = \pi(D - 2c) \tan \theta \tag{6.2}$$

$$W = \frac{\pi(D - 2c) \sin \theta}{m} - e \tag{6.3}$$

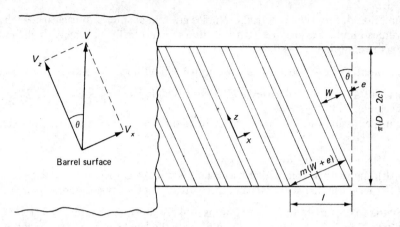

Figure 6.2   *Channels unrolled on to the plane through the flight tips*

Also, for an axial length, L, of screw, the length along the helical channel is

$$Z = \frac{L}{\sin \theta} \qquad (6.4)$$

The cartesian coordinate axes x, y, z may be chosen as shown in figures 6.2 and 6.3, the latter being a view along a screw channel in the downstream direction parallel to the flights. Figure 6.2 also shows a portion of the barrel surface moving with velocity V relative to the screw, which may be resolved into components V_z and V_x in the downstream and transverse (normal to the flights) directions

$$V = \pi DN, \; V_z = V \cos \theta, \; V_x = V \sin \theta \qquad (6.5)$$

Turning to the conservation equations governing melt flow, the arguments presented in sections 3.5 and 4.2.1 permit the following assumption.

(7) The density of the melt is locally constant.

Continuity equation 4.8 is then applicable. Of the terms in the general momentum-conservation equation 4.10, the body forces have already been eliminated by

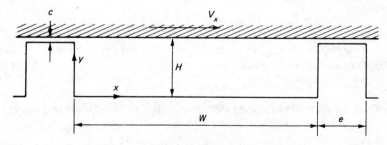

Figure 6.3   *View down a screw channel in the positive z direction*

assumption 4. From the practical examples given in chapter 5, it is clear that the Reynolds numbers associated with melt flows are normally extremely small, and a further assumption is possible.

(8) Inertia effects are negligible in comparison with viscous and pressure forces.

Equations 4.11, a typical cartesian form of which is given by equation 4.12, govern flow equilibrium.

Further assumptions normally made in connection with conservation of energy are as follows.

(9) The flow is steady.
(10) The thermal conductivity, k, is locally constant.
(11) The specific heat at constant pressure, C_p, is locally constant.

Only the first of these three assumptions is likely to cause significant errors: unsteady flow is discussed in section 6.6 under the heading of surging. Equation 4.17 is the appropriate form of energy equation.

Because melts are subjected to large rates of deformation for relatively long times while flowing in extruder screw channels, it is reasonable to make the following assumption.

(12) Melts may be treated as inelastic viscous fluids.

For the reasons discussed in section 4.3, the most general constitutive equation for an inelastic fluid, namely the stokesian-fluid model, is impractical to use. Suppose that the following assumption is made.

(13) Cross-viscosity effects are unimportant and the third invariant of the rate-of-deformation tensor is small.

Empirical power-law equation 4.38, together with the form of temperature dependence displayed in equation 3.19, can then be used to relate stresses and rates of deformation. For extruder channel flows, the following is normally a reasonable assumption.

(14) Viscosity is independent of pressure.

From figures 6.2 and 6.3, the boundary conditions for velocity components u, v and w in channel coordinate directions x, y and z are

$$u = v = w = 0 \text{ on } x = 0, x = W, y = 0$$
$$u = V_x, v = 0, w = V_z \text{ on } y = H \tag{6.6}$$

because there is no flow into the channel boundaries and, as discussed in section 4.4, the following assumption is made.

(15) There is no slip at the boundaries.

At least for the purposes of solving the channel flow equations it is reasonable to make the following assumption.

(16) There is negligible leakage of melt over the flight tips.

Therefore, within a channel

$$\int_0^H u \, dy = 0 \qquad (6.7)$$

Thermal boundary conditions may involve both temperatures and temperature derivatives, as discussed in section 4.4. At the barrel surface, a reasonable assumption is as follows.

(17) There is good thermal contact between the melt and metal surfaces.

In other words

$$T = T_b(z) \text{ on } y = H \qquad (6.8)$$

where T_b is the temperature of the inner surface of the barrel. Similarly, at the screw surface

$$T = T_s(z) \text{ on } y = 0, x = 0, x = W \qquad (6.9)$$

where T_s is the temperature of the screw, which in practice is not normally known. Alternative conditions sometimes used are

$$\frac{\partial T}{\partial y} = 0 \text{ on } y = 0, \qquad \frac{\partial T}{\partial x} = 0 \text{ on } x = 0, x = W \qquad (6.10)$$

which assume that a negligible amount of heat is conducted between the melt and screw. At least for an uncooled screw, this would appear to be a reasonable assumption. The best choice of temperature boundary condition for the screw is, however, intimately linked with the particular choice of mathematical model for melt flow, and is discussed further in section 6.1.4. The thermal convection terms in the energy equation allow for development of temperature in the direction of flow and hence, if these terms are to be retained, it is necessary to specify some initial temperature profile at the beginning of the region where melt flow is to be analysed.

Further simplifications to the channel flow equations may be introduced with the aid of the lubrication approximation, which was discussed in section 4.6. As the screw channel depth is either constant or varies slowly in the downstream direction, it is reasonable to apply the lubrication approximation in the z direction to velocities and to make the following assumption.

(18) The velocities are fully developed in the downstream direction.

In other words

$$u = u(x, y), \qquad v = v(x, y), \qquad w = w(x, y) \qquad (6.11)$$

Continuity equation 4.8 becomes

$$\frac{\partial u}{\partial x} + \frac{\partial v}{\partial y} = 0 \qquad (6.12)$$

and equilibrium equations 4.11 become

$$\frac{\partial p}{\partial x} = \frac{\partial \tau_{xx}}{\partial x} + \frac{\partial \tau_{xy}}{\partial y} \qquad (6.13)$$

$$\frac{\partial p}{\partial y} = \frac{\partial \tau_{yx}}{\partial x} + \frac{\partial \tau_{yy}}{\partial y} \tag{6.14}$$

$$\frac{\partial p}{\partial z} = P_z = \frac{\partial \tau_{zx}}{\partial x} + \frac{\partial \tau_{zy}}{\partial y} \tag{6.15}$$

where the pressure gradient P_z is independent of x and y.

For the reasons discussed in section 4.6, it is much less reasonable to apply the lubrication approximation in the z direction to temperatures. However, suppose that the following assumption is made.

(19) The temperatures are fully developed in the downstream direction.

Then $T = T(x, y)$, and energy equation 4.17 reduces to

$$\rho C_p \left(u \frac{\partial T}{\partial x} + v \frac{\partial T}{\partial y} \right) = k \left(\frac{\partial^2 T}{\partial x^2} + \frac{\partial^2 T}{\partial y^2} \right) + 4\mu I_2 \tag{6.16}$$

where

$$I_2 = \frac{1}{2} \left(\frac{\partial u}{\partial x} \right)^2 + \frac{1}{2} \left(\frac{\partial v}{\partial y} \right)^2 + \frac{1}{4} \left(\frac{\partial u}{\partial y} + \frac{\partial v}{\partial x} \right)^2$$

$$+ \frac{1}{4} \left(\frac{\partial w}{\partial y} \right)^2 + \frac{1}{4} \left(\frac{\partial w}{\partial x} \right)^2 \tag{6.17}$$

The next possible stage of simplification is to apply the lubrication approximation in the x direction. Assumption 2 ensures that the channel depth is constant in this direction, except at the flights. Therefore, for the approximation to be valid, the following assumption is necessary.

(20) The influence of the flights is negligible, and the flow may be treated as though the channel were infinitely wide.

Symbolically

$$H \ll W \tag{6.18}$$

and equations 6.11 and 6.13–6.17 reduce to

$$u = u(y), \ v = 0, \ w = w(y) \tag{6.19}$$

$$\frac{\partial p}{\partial x} = P_x = \frac{d\tau_{xy}}{dy} \tag{6.20}$$

$$\frac{\partial p}{\partial y} = 0 \tag{6.21}$$

$$P_z = \frac{d\tau_{zy}}{dy} \tag{6.22}$$

$$0 = k \frac{\partial^2 T}{\partial y^2} + 4\mu I_2 \tag{6.23}$$

$$I_2 = \frac{1}{4}\left(\frac{du}{dy}\right)^2 + \frac{1}{4}\left(\frac{dw}{dy}\right)^2 \tag{6.24}$$

where the pressure gradient P_x is independent of x and y. Equation 6.12 is automatically satisfied, although continuity must still be satisfied in terms of over-all channel flow rate. The flow is now two dimensional in the sense that there are only two non-zero velocity components, u and w, which are functions of the third coordinate y.

Another very useful combination of assumptions is provided by taking velocity profiles to be locally fully developed in both the downstream and transverse directions, and retaining thermal convection in the downstream direction only. The energy equation becomes

$$\rho C_p \frac{\partial T}{\partial z} = k \frac{\partial^2 T}{\partial y^2} + 4\mu I_2 \tag{6.25}$$

where I_2 is given by equation 6.24. This mathematical model gives rise to the developing flow solution described in section 6.1.4.

Further simplification of equations 6.20–6.24 involves the following assumption.

(21) The contribution of the transverse flow to viscosity determination and viscous heating is negligible.

Equation 6.20 is no longer required, and equation 6.24 reduces to

$$I_2 = \frac{1}{4}\left(\frac{dw}{dy}\right)^2 \tag{6.26}$$

The flow is now one dimensional.

Two final assumptions that are sometimes applied to one-dimensional, two-dimensional or more sophisticated models are the isothermal assumption, which was discussed in section 4.5, and the assumption that the melt is newtonian and its viscosity constant. If the latter assumption is made, a reasonable value for this viscosity is that defined at some suitable temperature and shear rate, such as the barrel temperature and mean downstream shear rate, V_z/H.

6.1.1 Dimensionless Parameters for Screw Channel Melt Flow

The parameters derived in section 4.5 from a dimensional analysis of a typical channel flow may also be defined for screw-channel flow. The characteristic flow velocity may be taken as V_z, the downstream component of the barrel velocity relative to the screw. A case could be made for using V, the resultant circumferential velocity, although V_z is preferred here because it is in the direction of the volumetric flow rate, Q

$$Q = \int_0^W \int_0^H w \, dy \, dx \tag{6.27}$$

Therefore, the Reynolds number, dimensionless pressure gradient, Peclet, Griffith and Graetz numbers previously defined in equations 4.60–4.65 become

$$Re = \frac{\rho V_z H}{\bar{\mu}} \tag{6.28}$$

$$\pi_P = \frac{P_z H}{\bar{\tau}} \tag{6.29}$$

$$Pe = \frac{\rho C_p V_z H}{k} \tag{6.30}$$

$$G = \frac{b\bar{\tau}\bar{\gamma}H^2}{k} \tag{6.31}$$

$$Gz = \frac{\rho C_p V_z H^2}{kZ} \tag{6.32}$$

where Z is the helical length defined according to equation 6.4. The mean viscosity and shear stress, $\bar{\mu}$ and $\bar{\tau}$, are determined at the mean shear rate, $\bar{\gamma} = V_z/H$, and barrel temperature, T_b. Hence, from equations 4.38 and 3.19

$$\bar{\mu} = \mu_0 \left(\frac{\bar{\gamma}}{\gamma_0}\right)^{n-1} \exp\left[-b(T_b - T_0)\right], \bar{\tau} = \bar{\mu}\bar{\gamma} \tag{6.33}$$

A further useful parameter is the dimensionless flow rate

$$\pi_Q = \frac{Q}{WHV_z} \tag{6.34}$$

which expresses the ratio between the actual flow rate along the channel and the flow rate that would have been achieved if all the melt moved downstream with the same velocity as the barrel relative to the screw. It is therefore a measure of volumetric efficiency.

The following practical example illustrates the calculation of the above dimensionless parameters. A typical polystyrene melt has the following properties

$$\mu_0 = 10.8 \text{ kN s/m}^2 \text{ at } \gamma_0 = 1 \text{ s}^{-1} \text{ and } T_0 = 200 \text{ °C}, n = 0.36$$

$$b = 0.022 \text{ °C}^{-1}, \rho = 990 \text{ kg/m}^3, C_p = 2000 \text{ J/kg °C}$$

$$k = 0.21 \text{ W/m °C}$$

Suppose this material flows at a rate of 510 kg/h along the metering section of a screw rotating at $N = 100$ rev/min, the barrel being maintained at $T_b = 220$ °C. The barrel diameter is $D = 120$ mm, axial length of metering section $L = 960$ mm, screw-channel depth $H = 6$ mm and flight width $e = 12$ mm. The screw is single start with flight pitch equal to the diameter.

Assuming that the clearance between the screw and barrel is small compared with the diameter, equations 6.2 to 6.5 may be used to derive the helix angle, channel width, helical length and downstream component of the barrel relative velocity

$$0 = \tan^{-1}\left(\frac{l}{\pi D}\right) = \tan^{-1}\left(\frac{1}{\pi}\right) = 17.66°$$

$$W = \pi D \sin 0 \quad e = \pi \times 120 \times 0.3033 - 12 = 102 \text{ mm}$$

$$Z = \frac{L}{\sin 0} = \frac{0.96}{0.3033} = 3.17 \text{ m}$$

$$V_z = \pi D N \cos 0 = \frac{\pi \times 120 \times 100 \times 0.9529}{60} = 599 \text{ mm/s}$$

The mean shear rate is $\bar{\gamma} = V_z/H = 99.8 \text{ s}^{-1}$ and, from equations 6.33, the mean viscosity and shear stress are

$$\bar{\mu} = 10.8 \times 99.8^{-0.64} \times \exp\left[-0.022(220 - 200)\right] = 0.366 \text{ kN s/m}$$

$$\bar{\tau} = \bar{\mu}\bar{\gamma} = 0.366 \times 99.8 = 36.5 \text{ kN/m}^2$$

The characteristic Reynolds, Peclet, Griffith and Graetz numbers for the flow are

$$Re = \frac{\rho V_z H}{\bar{\mu}} = \frac{990 \times 599 \times 10^{-3} \times 6 \times 10^{-3}}{0.366 \times 10^3} = 9.72 \times 10^{-3}$$

$$Pe = \frac{\rho C_p V_z H}{k} = \frac{990 \times 2000 \times 599 \times 10^{-3} \times 6 \times 10^{-3}}{0.21} = 3.39 \times 10^4$$

$$G = \frac{b\bar{\tau}\bar{\gamma}H^2}{k} = \frac{0.022 \times 36.5 \times 10^3 \times 99.8 \times (6 \times 10^{-3})^2}{0.21} = 13.7$$

$$Gz = Pe\frac{H}{Z} = 3.39 \times 10^4 \times \frac{6 \times 10^{-3}}{3.17} = 64.2$$

The typically small Reynolds number confirms that inertia effects are negligible in screw channel flows. The large size of the Griffith number implies that the flow is far from being isothermal, and the temperatures generated significantly affect melt viscosity. This is in contrast to many of the die flows considered in chapter 5, where the Griffith number was relatively small. With large values of Pe and Gz, thermal convection is the dominant mode of heat transfer. Note that, because the Graetz number is substantially greater than 10, thermally fully developed flow is not achieved even at the end of the metering section. Thus, by merely calculating the characteristic dimensionless flow parameters, a considerable amount of insight into the nature of the flow can be gained.

Two further parameters not so far calculated are the dimensionless flow rate and pressure gradient. The volumetric flow rate can be found from the mass flow rate as

$$Q = \frac{510}{990} = 0.515 \text{ m}^3/\text{h} = 1.43 \times 10^{-4} \text{ m}^3/\text{s}$$

and the dimensionless flow rate as

$$\pi_Q = \frac{Q}{WHV_z} = \frac{1.43 \times 10^{-4}}{102 \times 10^{-3} \times 6 \times 10^{-3} \times 599 \times 10^{-3}} = 0.390$$

If the pressure gradient is known, the value of π_P can also be calculated. In practice, however, this is only the case for machines specially equipped for measuring pressure profiles, normally those used for research and development purposes. Indeed, one of the objects of analysing melt flow is to predict pressure gradients, and this is often conveniently done in terms of dimensionless pressure gradients.

6.1.2 Newtonian Solutions

Although the use of the newtonian flow assumption of constant viscosity does not give quantitatively reliable solutions to the flow equations, they do demonstrate many of the important features of screw channel flow, and can be obtained by relatively simple analyses. For example, with constant melt viscosity, μ, equation 6.22 for equilibrium in the downstream direction of flow in a wide channel becomes

$$\frac{d^2 w}{dy^2} = \frac{P_z}{\mu} \tag{6.35}$$

Repeated integration and application of the boundary conditions defined by equations 6.6 gives the velocity profile

$$w = y \frac{V_z}{H} - \frac{y(H - y)}{2} \frac{P_z}{\mu} \tag{6.36}$$

Note that the two terms on the right-hand side of this equation can be identified as a drag-flow component due to velocity V_z, and a pressure flow component due to gradient P_z: for a constant viscosity material, drag and pressure flows can be super-imposed. Figures 6.4a and b show the drag and pressure flows, respectively, the

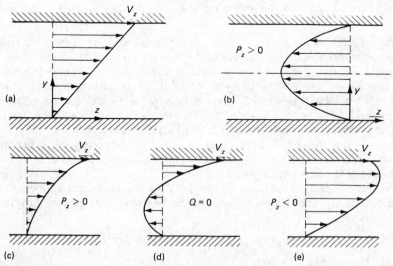

Figure 6.4   Downstream velocity profiles for newtonian melt flow; (a) pure drag flow; (b) pressure flow; (c)–(e) drag and pressure flows combined in varying proportions

latter for the case of a positive pressure gradient, while figures 6.4c–e show these simple flows combined in varying proportions. Figure 6.4c shows the case of a relatively small positive pressure gradient, that is, when the pressure increases gradually along the screw towards the die. Figure 6.4d shows the extreme form of this condition, when the pressure generating capacity of the screw is a maximum and there is no downstream flow rate. Figure 6.4e shows the situation where the pressure is decreasing along the screw towards the die, which may occur in the metering section.

The downstream velocity profile given by equation 6.36 may be integrated according to equation 6.27 to give the volumetric flow rate along the channel

$$Q = \frac{WHV_z}{2} - \frac{WH^3 P_z}{12\mu} \tag{6.37}$$

The dimensionless form of this result may be obtained, using the definitions given in equations 6.29 and 6.34, as

$$\pi_Q = \frac{1}{2} - \frac{\pi_P}{12} \tag{6.38}$$

The transverse flow in the x direction is governed by equation 6.20, which for a melt of constant viscosity becomes

$$\frac{d^2 u}{dy^2} = \frac{P_x}{\mu} \tag{6.39}$$

Integration and the boundary conditions of equations 6.6 give the velocity profile

$$u = y\frac{V_x}{H} - \frac{y(H-y)}{2}\frac{P_x}{\mu} \tag{6.40}$$

Now, in order to satisfy the condition of zero transverse flow in equation 6.7

$$P_x = \frac{6\mu V_x}{H^2} \tag{6.41}$$

and

$$u = V_x \left[3\left(\frac{y}{H}\right)^2 - 2\left(\frac{y}{H}\right) \right] \tag{6.42}$$

The shape and form of this profile in the x direction is as shown in figure 6.4d for zero downstream flow rate. Figure 6.5 shows typical combined downstream and transverse velocity profiles. The flow recirculates in the transverse plane, and the effect of the superimposed downstream flow is to cause the melt to move in a flattened helix along the screw channel. Note that, when the downstream pressure gradient is large and positive, the downstream velocity component becomes negative near the screw root. This does not mean, however, that there is backflow in the direction parallel to the screw axis. In the extreme case of $Q = 0$, the downstream and transverse velocity profiles combine to give flow along closed paths in the plane normal to the screw axis.

The above solutions for downstream and transverse screw channel flow are for

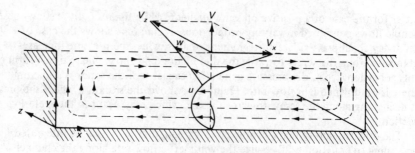

Figure 6.5   *Combined downstream and transverse channel flows*

an infinitely wide channel, that is, one in which the presence of the flights affects
the velocity profiles over a negligible proportion of the total channel width. If the
condition expressed in equation 6.18 is relaxed, and the channel assumed to be of
finite width, downstream flow is governed by equation 6.15, which for a newtonian
melt becomes

$$\frac{\partial^2 w}{\partial x^2} + \frac{\partial^2 w}{\partial y^2} = \frac{P_z}{\mu} \tag{6.43}$$

A series solution for this equation with the appropriate boundary conditions can be
expressed in a form very similar to equation 6.36

$$w = y \frac{V_z}{H} f_D - \frac{y(H-y)}{2} \frac{P_z}{\mu} f_P \tag{6.44}$$

where the drag and pressure flow velocity factors, f_D and f_P, are convergent infinite
series, the values of which both tend to one as $H/W \to 0$. The detailed forms of these
series are less important than those of the corresponding factors associated with
flow rate. If equation 6.44 is integrated according to equation 6.27, and the
resulting flow rate equation expressed in dimensionless form, a generalised form of
equation 6.38 is obtained as

$$\pi_Q = \frac{F_D}{2} - \frac{\pi_P F_P}{12} \tag{6.45}$$

F_D and F_P are the drag and pressure flow shape factors, respectively, and are the
following functions of the channel shape ratio H/W

$$F_D = \frac{16}{\pi^3} \frac{W}{H} \sum_{g=1,3,5,\ldots}^{\infty} \frac{1}{g^3} \tanh \left(\frac{g\pi}{2} \frac{H}{W} \right) \tag{6.46}$$

$$F_P = 1 - \frac{192}{\pi^5} \frac{H}{W} \sum_{g=1,3,5,\ldots}^{\infty} \frac{1}{g^5} \tanh \left(\frac{g\pi}{2} \frac{W}{H} \right) \tag{6.47}$$

Some typical values are given in table 6.1. Also, in the practical example considered
in section 6.1.1, $H/W = 0.0587$ and $F_D = 0.974$, $F_P = 0.963$. Therefore, the presence

TABLE 6.1
Drag and pressure flow shape factors

H/W	F_D	F_P
0	1.0	1.0
0.1	0.946	0.937
0.2	0.891	0.874
0.3	0.837	0.811

of the flight walls at the sides of the channel in the metering section of a typical screw affects the calculated flow rate by only a few per cent, and the application of the lubrication approximation in the transverse channel direction is normally reasonable, at least as far as downstream flow rate is concerned.

Further refinements to the newtonian solutions are, of course, possible. For example, a solution could be obtained for transverse flow in a channel of finite width, and both downstream and transverse flow could be made nonisothermal. Such changes are not, however, justified until the newtonian flow assumption is relaxed to permit the treatment of melt viscosity as a function of the local rates of deformation.

6.1.3 Non-newtonian Solutions

The simplest form of solution corresponding to non-newtonian screw channel flow is that for one-dimensional isothermal downstream flow. Equation 6.22 governs the flow

$$\frac{d\tau}{dy} = P_z \tag{6.48}$$

where $\tau \equiv \tau_{zy}$, and from constitutive equation 4.38

$$\tau = \mu_0 \left| \frac{\gamma}{\gamma_0} \right|^{n-1} \gamma, \qquad \gamma = \frac{dw}{dy} \tag{6.49}$$

Integration to find the downstream velocity profile is complicated by the analytical difficulty first mentioned in section 5.1.1, and encountered more seriously in section 5.1.3. Using the empirical power-law constitutive equation, the change in sign of the shear stress and shear rate at the stress neutral surface is difficult to deal with when the position of this surface is not known in advance.

Following the procedure adopted in section 5.1.2 — that is, choosing the origin for the coordinate normal to the flow boundaries to be at the stress neutral surface — one way to proceed with the present analysis is to define a new coordinate y', measured from the $\tau = 0$ position as shown in figure 6.6. This position is defined as λH above the screw surface, where λ can take any real value between $-\infty$ and $+\infty$, the extremes corresponding to pure drag flow, and $\lambda = 1/2$ to pressure flow.

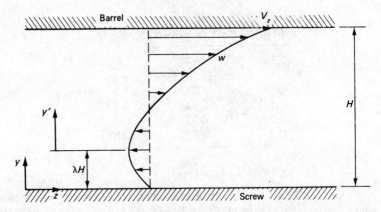

Figure 6.6   *Non-newtonian downstream melt flow showing new coordinate measured from the stress neutral surface*

Equation 6.48 can be integrated to give the linear shear-stress profile

$$\tau = P_z y' \tag{6.50}$$

Introducing the dimensionless pressure gradient defined by equation 6.29

$$\frac{\tau}{\bar{\tau}} = \pi_P \frac{y'}{H}$$

Hence, using equation 6.49

$$\frac{\gamma}{\bar{\gamma}} = \left| \frac{\tau}{\bar{\tau}} \right|^{1/n} \operatorname{sgn}(\tau) = \left| \pi_P \frac{y'}{H} \right|^{1/n} \operatorname{sgn}(\pi_P y')$$

and integration, together with the boundary condition $w = 0$ at $y' = -\lambda H$, yields the downstream velocity

$$w = V_z \, |\pi_P|^{1/n} \operatorname{sgn}(\pi_P) \frac{n}{n+1} \left(\left| \frac{y'}{H} \right|^{1/n+1} - |\lambda|^{1/n+1} \right)$$

The second boundary condition, $w = V_z$ at $y' = H(1 - \lambda)$, gives the following equation for λ

$$\frac{n+1}{n} = |\pi_P|^{1/n} \operatorname{sgn}(\pi_P)(|1 - \lambda|^{1/n+1} - |\lambda|^{1/n+1}) \tag{6.51}$$

The volumetric flow rate per unit width of channel is

$$\frac{Q}{W} = \int_{-\lambda H}^{H(1-\lambda)} w \, dy' = V_z \, |\pi_P|^{1/n} \operatorname{sgn}(\pi_P) \frac{n}{n+1} \left[\frac{y'n}{2n+1} \left| \frac{y'}{H} \right|^{1/n+1} \right.$$

$$\left. - y' |\lambda|^{1/n+1} \right]_{-\lambda H}^{H(1-\lambda)}$$

from which may be derived, with the aid of equation 6.51, the following expression for the dimensionless flow rate

$$\pi_Q = \frac{Q}{WHV_z} = 1 - \frac{n\lambda}{2n+1} - \frac{n}{2n+1} \mid \pi_P \mid^{1/n} \text{sgn} \,(\pi_P) \mid 1 - \lambda \mid^{1/n+1} \quad (6.52)$$

Equations 6.51 and 6.52 contain four variables: n, λ, π_P and π_Q. Assuming that the material parameter n is known, λ can in principle be eliminated to give the required relationship between π_Q and π_P. Note that for a non-newtonian melt it is no longer possible to separate and superimpose drag and pressure flows.

Further analytical progress is only possible for particular values of n. For example, for the newtonian case, $n = 1$, equation 6.51 gives

$$\lambda = \frac{1}{2} - \frac{1}{\pi_P} \quad (6.53)$$

and π_Q is related to π_P by equation 6.38. The other special case is $n = 1/2$, for which equations 6.51 and 6.52 become

$$3 = \pi_P^2 \, \text{sgn} \, (\pi_P)(\mid 1 - \lambda \mid^3 - \mid \lambda \mid^3) \quad (6.54)$$

$$\pi_Q = 1 - \frac{\lambda}{4} - \frac{\pi_P^2}{4} \, \text{sgn} \, (\pi_P) \mid 1 - \lambda \mid^3 \quad (6.55)$$

Now, provided the stress neutral surface does not lie within the flow — that is, $\lambda \leqslant 0$ or $\lambda \geqslant 1$ — equation 6.54 can be simplified to give

$$\lambda = \frac{1}{2} - \frac{1}{\pi_P} \left(1 - \frac{\pi_P^2}{12} \right)^{1/2} \quad (6.56)$$

and equation 6.55 becomes

$$\pi_Q = \frac{1}{2} - \frac{\pi_P}{6} \left(1 - \frac{\pi_P^2}{12} \right)^{1/2} \quad (6.57)$$

The restrictions in terms of λ on the validity of these results can also be expressed as either $\pi_P^2 \leqslant 3$, or $1/4 \leqslant \pi_Q \leqslant 3/4$. Note again how the position of the stress neutral surface affects the analytical solubility of the problem.

In general, the elimination of λ from equations 6.51 and 6.52 cannot be performed analytically, although it is a straightforward numerical problem. The simplest approach is to obtain π_P and π_Q from equations 6.51 and 6.52, respectively, for a series of values of λ covering the range of interest, and so produce a flow curve of the form shown in figure 6.7. Alternatively, π_Q can be obtained numerically for a given π_P, or vice versa. The straight line shown for $n = 1$ in figure 6.7 represents equation 6.38, and part of the curve for $n = 1/2$ is given by equation 6.57. All the curves for different values of n pass through $\pi_Q = 1/2$ under drag flow conditions, because under such conditions the downstream velocity profile is linear, the shear rate is constant and non-newtonian material behaviour is not important. In the presence of significant pressure gradients of either sign, however, it is clear from the flow curves that the use of a newtonian analysis seriously over-estimates the magnitudes of such gradients.

In the case of non-newtonian flow, the inclusion of transverse flow to make the

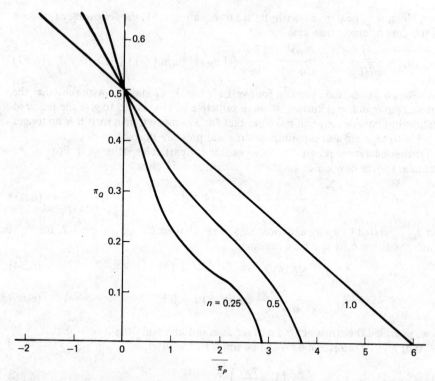

Figure 6.7   *Dimensionless flow rate plotted against dimensionless pressure gradient for one-dimensional isothermal power-law flow*

solutions two dimensional increases considerably the complexity of the analysis. Equations 6.20 and 6.22 must now be solved simultaneously with the aid of equations 4.38, 6.24, 6.6 and 6.7. Results may be expressed in the form of figure 6.7, the effect of the inclusion of transverse flow being to displace the characteristics by relatively small amounts. The curves no longer pass through a common point for zero π_P, and have the helix angle as an additional independent variable. Still retaining the isothermal assumption, the effect of the screw flights in forming a channel of finite width can be examined. For typical screw channels, this effect proves to be relatively small, smaller even than for newtonian flow.

Because fully three-dimensional solutions for melt flow in a screw channel are difficult to justify in terms of the cost of digital computing resources required, further refinements of the non-newtonian flow analysis tend to progress in one of two directions. Either velocities and temperatures can be assumed to be fully developed at a particular channel cross section, or the flow can be treated as a two-dimensional one developing in the downstream direction. In view of the conclusions reached in the practical example on the length of channel required for thermal development (section 6.1.1), it is perhaps not surprising that the latter approach generally proves to be the more useful.

The most complete form of solution for flow that is fully developed in the downstream direction allows for the finite width of the channel. The governing equations are equations 6.12–6.17, 4.38 and 3.19, subject to the boundary conditions given by equations 6.6, 6.8 and 6.9 or 6.10. A set of four simultaneous coupled second-order nonlinear partial differential equations may be derived, two governing transverse flow (obtained from equations 6.12 to 6.17), and the other two governing downstream flow and heat transfer (equations 6.15 and 6.16). This problem can only be solved numerically, and has been studied extensively using both finite difference and finite element methods. An important conclusion is that convection associated with the recirculating transverse flow provides the main mechanism for heat transfer between barrel and screw. Another is that, for typical screw channels with relatively small H/W, it is reasonable to apply the lubrication approximation in the transverse direction (assumption 20), provided a temperature boundary condition for the screw that accounts for convection in the transverse flow is used.

Consideration of a dimensionless form of the energy conservation equation – such as equation 4.58, derived by dimensional analysis of a typical melt flow in section 4.5 – leads to an appropriate choice of thermal boundary condition. If the Peclet number is very large, which it is in most extruder channel flows, then, in order to balance the thermal convection terms with the conduction and dissipation terms, the products of velocities and temperature gradients in the directions of these velocities must be very small. In other words, temperature gradients in directions of flow – that is, along streamlines – must be very small. Isotherms therefore tend to take up shapes that lie along streamlines. Applying this argument to recirculating transverse screw channel flow, it follows that, since the screw surface is a streamline, it must also be very nearly an isotherm. This is consistent with the fact that the metallic screw is a much better conductor of heat than the polymer melt, and its surface is therefore at an essentially constant temperature. Neglecting the small amount of melt that may flow over the tips of the screw flights, the screw surface also forms the same streamline as the barrel surface, namely the one completely enclosing the recirculating flow. It is therefore to be anticipated that the screw temperature is equal to the barrel temperature. This is shown to be the case in practice by experimental measurements of temperatures in screw metering sections. A suitable form of the temperature boundary condition for the screw given by equation 6.9 is

$$T = T_b(z) \text{ on } y = 0, x = 0, x = W \qquad (6.58)$$

In the limiting case of infinite Peclet number, this also corresponds to the zero conduction heat transfer conditions expressed in equations 6.10.

6.1.4 Developing Flow Solution

The following mathematical model for developing non-newtonian melt flow in an extruder screw channel offers one of the best compromises for routine use in screw design work, in terms of its ability to predict actual melt flow behaviour for an acceptable amount of numerical computation. Because the complexity of the model demands the services of a digital computer, it is not possible to express the results as

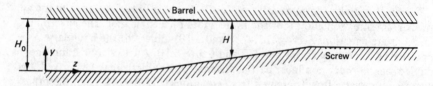

Figure 6.8   *Cross-section of an unwrapped screw channel in the downstream direction*

algebraic formulae: the result is the computer program, which allows the solution to any particular problem to be obtained.

Figure 6.8 shows the cross section of a typical unwrapped screw channel in the y, z plane, coordinate directions that were shown previously in figures 6.2 and 6.3. In general, it may be necessary to consider melt flow not only in the metering section of the screw, but also in the feed and compression sections. This is particularly true of melt-fed extruders used as homogenisers. Note that, in the analysis that follows, the radial coordinate, y, is assumed to have its origin always at the screw root. Thanks to the slowly varying channel depth, which is such that

$$\left| \frac{\partial H}{\partial z} \right| \ll 1 \tag{6.59}$$

the resulting distortion of the truly cartesian coordinate system implied by figure 6.8 is not significant.

Also by virtue of inequality 6.59, the geometric conditions necessary for the lubrication approximation to be applicable in the downstream direction are satisfied. Following the argument in section 4.6, velocity profiles at a particular downstream position may be assumed to be locally fully developed, but this is not true of temperature profiles. Velocity profiles may change slowly with downstream position, as a result of the changing temperature profile. Also, following the conclusions of the last subsection, the lubrication approximation is applied in the transverse direction to both velocities and temperatures. Hence, for the purposes of analysis

$$u = u(y), \qquad v = 0, \qquad w = w(y), \qquad T = T(y, z) \tag{6.60}$$

The equations of equilibrium for the transverse and downstream directions are equations 6.20 and 6.22, respectively, and energy equation 4.17 reduces to

$$\rho C_p w \frac{\partial T}{\partial z} = k \frac{\partial^2 T}{\partial y^2} + 4\mu I_2 \tag{6.61}$$

where I_2 is given by equation 6.24. The argument for neglecting thermal conduction in the downstream direction, while retaining that in the radial direction, was given in section 4.5. Velocity and temperature boundary conditions are given by equations 6.6, 6.8 and 6.58, while equation 6.7 imposes the required recirculation on the transverse flow. An initial condition for the temperature at $z = 0$, the start of the flow, is also required. Finally, over-all conservation of mass is preserved by requiring the volumetric flow rate

$$Q = W \int_0^H w \, dy \tag{6.62}$$

to be independent of downstream position.

Introducing the following dimensionless variables and relative channel depth, S

$$W^* = \frac{w}{V_z}, \qquad U^* = \frac{u}{V_z}, \qquad Y^* = \frac{y}{H}, \qquad Z^* = \frac{z}{H_0}$$

$$S = \frac{H}{H_0}, \qquad T^* = b[T - T_b(0)] \tag{6.63}$$

where H_0 and $T_b(0)$ are, respectively, the channel depth and barrel temperature at $z = 0$, the equilibrium and energy equations become, using constitutive equations 4.38 and 3.19

$$\pi_X = \frac{d}{dY^*}\left[\frac{dU^*}{dY^*}(4I_2^*)^{(n-1)/2}\exp(-T^*)\right] \tag{6.64}$$

$$\pi_P = \frac{d}{dY^*}\left[\frac{dW^*}{dY^*}(4I_2^*)^{(n-1)/2}\exp(-T^*)\right] \tag{6.65}$$

$$S^2\, Pe W^* \frac{\partial T^*}{\partial Z^*} = \frac{\partial^2 T^*}{\partial Y^{*2}} + GS^{1-n}(4I_2^*)^{(n+1)/2}\exp(-T^*) \tag{6.66}$$

The dimensionless form of I_2 is

$$I_2^* = \frac{1}{4}\left(\frac{dW^*}{dY^*}\right)^2 + \frac{1}{4}\left(\frac{dU^*}{dY^*}\right)^2 = \left(\frac{H}{V_z}\right)^2 I_2 \tag{6.67}$$

The dimensionless downstream pressure gradient, π_P, is as defined in equation 6.29 in terms of local channel depth and local mean shear stress. Similarly, the dimensionless transverse pressure gradient, π_X, is

$$\pi_X = \frac{P_x H}{\bar{\tau}} \tag{6.68}$$

where $\bar{\tau}$ is given by equations 6.33. On the other hand, the Peclet and Griffith numbers appearing in equation 6.66 are as defined by equations 6.30 and 6.31 in terms of the channel depth, H_0, mean shear rate, V_z/H_0, and mean shear stress at the channel inlet, $z = Z^* = 0$. The boundary conditions for the dimensionless velocities are

$$U^* = W^* = 0 \text{ at } Y^* = 0$$

$$U^* = \tan\theta, \qquad W^* = 1 \text{ at } Y^* = 1 \tag{6.69}$$

and the following integral conditions must be satisfied

$$\int_0^1 U^* \, dY^* = 0, \qquad \int_0^1 W^* \, dY^* = \pi_Q = \frac{Q}{WHV_z} \tag{6.70}$$

A numerical method of solution may be outlined as follows. Equations 6.64 and 6.65 may be integrated subject to boundary conditions 6.69 to find the dimensionless velocity profiles at the channel inlet, where the initial temperature profile is known. Equation 6.66 expressed in finite difference form may then be used to step a small distance downstream to find the temperature profile there. Repetition of the velocity analysis completes the calculations at the new channel section, and makes it possible to take a further downstream step in terms of temperatures, and so on until the end of the channel is reached. The velocity and temperature analyses are described in detail in appendix B.

In order to illustrate the use of the developing flow model, consider again the example described in section 6.1.1, concerning polystyrene melt flow in the metering section of a 120 mm extruder. Figure 6.9 shows the predicted variations of dimensionless pressure gradient, pressure and bulk mean temperature along the helical length of the screw channel. The melt is assumed to enter at a uniform temperature of 220 °C, the set barrel temperature. The bulk mean temperature,

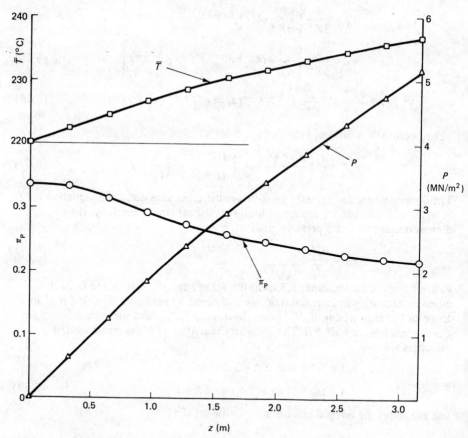

Figure 6.9   *Variations of dimensionless pressure gradient, pressure and bulk mean melt temperature with downstream position*

which may be defined as

$$\overline{T} = \frac{1}{Q} \int_0^H wT \, dy \qquad\qquad (6.71)$$

provides a convenient means of characterising the temperature profile at any downstream position. It is the uniform temperature that would be obtained if the melt were collected in a perfectly insulated container and allowed to reach thermal equilibrium.

Figure 6.9 shows that, whereas with a fully developed flow solution there is a unique relationship between π_P and π_Q, such as the one shown in figure 6.7, with a developing flow π_P varies for constant π_Q. In this particular example, for a π_Q of 0.390, π_P decreases by some 36 per cent – in fact, from 0.336 to 0.215 – over the length of the channel, because of the increasing temperatures within the flow. The corresponding reduction in actual pressure gradient gives rise to the somewhat nonlinear pressure profile shown. The bulk mean temperature increases from 220 °C at the channel inlet to nearly 237 °C at the delivery end. The rise is seen to be smooth, and by no means complete. In other words, the flow at the delivery end is still far from being thermally fully developed, a conclusion previously established from a consideration of the Graetz number.

Figure 6.10 shows the predicted temperature profiles through the channel depth at positions halfway along, and at the delivery end of, the metering section. Note how in the relatively early stages of development the pattern of viscous dissipation creates two peaks in the temperature profile, which are subsequently smoothed out by conduction. Perhaps the most important practical conclusion to be drawn from figures 6.9 and 6.10, however, is that, despite controlling the barrel temperature, temperatures within the melt can be substantially higher – in this case, by up to 17 °C on average and 22 °C locally. This phenomenon, which becomes more

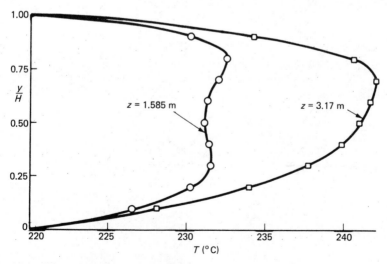

Figure 6.10   *Melt temperature profiles*

pronounced with increasing machine size, is not always appreciated in practice, because it is difficult to make melt temperature measurements that are not affected by the presence of the metal flow boundaries. A further application of the developing flow solution is provided by the design study outlined in section 6.7.4.

6.1.5 Flow in the Clearance

So far, attention has been confined to melt flow in the channel of an extruder screw: assumption 16 specifically neglected leakage of melt over the flight tips. While this proves to be reasonable for the purposes of channel flow analysis, some melt may flow in the small clearance between the barrel and screw flight. If there is no slip between the melt and metal surfaces, such leakage takes the form of a drag flow. Following the argument developed in section 5.5, the pressure gradients and differences necessary to affect the velocity profile significantly in such a thin film do not exist in practice.

In order to estimate the effects of leakage, the screw is assumed to run concentrically in the barrel, with radial clearance c. Figure 6.11 shows part of a screw unrolled on to the plane through the flight tips as in figure 6.2. The net volumetric flow rate across section AB is $\dot{m}/\rho$, where $\dot{m}$ is the extrudate mass flow rate. The component flow rates across AB are as follows: Q down each screw channel (in figure 6.11 two such channels are shown, $m = 2$); Q_F above each flight

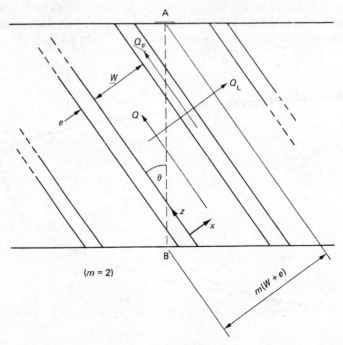

Figure 6.11   *Part of an unrolled screw with component flow rates shown*

in the downstream direction; and Q_L, the leakage flow rate normal to the flights. Hence

$$\frac{\dot{m}}{\rho} = mQ + mQ_F - Q_L \qquad (6.72)$$

If the leakage takes the form of a drag flow with a linear variation of circumferential velocity between the barrel and flight-tip surfaces

$$Q_F = \tfrac{1}{2}ceV_z, \qquad \frac{dQ_L}{dz} = \tfrac{1}{2}cV_x \qquad (6.73)$$

Now, the relevant downstream distance for evaluating Q_L is the projection of the length AB on to the z direction, namely $m(W + e)\cot\theta$. Hence, substituting the expressions for Q_F and Q_L into equation 6.72, the dimensionless channel flow rate may be obtained as

$$\pi_Q = \frac{Q}{WHV_z} = \frac{\dot{m}}{\rho mWHV_z} + \frac{c}{2H} \qquad (6.74)$$

The actual flow rate per channel is greater than the apparent one calculated from the output rate, and π_Q is increased by an amount $c/2H$. Therefore, if the metering section of a screw is generating a delivery pressure, the effect of leakage is to reduce the generating capacity.

There is a limited amount of experimental evidence to suggest that melt does not always flow in the clearance of an extruder. For example, measurements of power consumption (which is discussed in section 6.4) sometimes fail to detect the substantial amounts of energy that calculations suggest are dissipated in the clearance. Also, metal-to-metal contact between screw and barrel can be detected. Such contact, presumably due to the absence of the lubricating melt film between them, appears to occur only at shear stresses higher than the critical values associated with the onset of slip, as discussed in sections 3.3.3 and 3.4.

6.2 Solids Conveying in Extruders

In comparison with the large amount of research on melt flow in screw extruders, relatively little attention has been given to the conveying of solid granules or powder from the feed hopper of a plasticating extruder to the position at which a significant amount of melting has occurred. There are several reasons for this. Solids conveying is usually a relatively minor part of the over-all extrusion process, and occurs only over the first few turns of the screw channel. It rarely controls directly the output from the machine, which is normally determined by the later melting and melt pumping processes. An analysis of the solids conveying process can therefore only provide estimates of the upper limit on the output capacity of the machine and the pressures that can be generated in the feed section of the screw.

A set of continuum mechanics equations can be established for solids conveying in screw channels; the resulting equations are similar to those governing melt flow. Even though such a continuum approach ignores the particulate nature of polymer feedstocks, it is still difficult to obtain useful solutions. Fortunately, for many practical purposes, it proves to be reasonable to make the further simplification of treating material movement as plug flow in the channel, ignoring the effects of stresses and deformations within the bulk of the solid.

6.2.1 Simple Plug Flow Analysis

Experimental observations of polymer-granule motion in model extruders with transparent barrels serve to support the validity of the plug-flow assumption, under which the downstream velocity component is independent of position in the channel cross section. The main restriction is that the channel must be full of compacted material, a condition that is normally satisfied very early in a plasticating extruder, since the conveying rate is limited by the later melting and metering processes.

Some of the assumptions made in melt-flow analyses are also relevant to solid-plug flow. The assumptions regarding screw geometry (assumptions 1–3), locally constant bulk density (assumption 7), negligible inertia effects (assumption 8), steady flow (assumption 9) no leakage over the flight tips (assumption 16) are all applicable. Of the remaining assumptions that might be relevant to solid flow, only those regarding the unrolling of the screw channels (assumption 6) and no slip at the flow boundaries (assumption 15) are definitely not applicable. The relatively deep channels in screw-feed sections cannot be unrolled for analytical purposes without due allowance for curvature. Conditions at the boundaries of the solid plug are assumed to be determined by a coulomb-friction mechanism between the material and the screw and barrel surfaces. Assumptions 4 and 5 concerning negligible gravity and centrifugal forces are also valid once significant stresses have been generated in the compacted solid plug.

One way of presenting analytical results for plug flow is in terms of the feed angle at which the plug moves relative to planes perpendicular to the screw axis. Figure 6.12a shows a portion of the solid plug in a screw channel; α is the feed angle measured at the barrel surface, and the arrow shows the direction of plug motion relative to the stationary barrel. Figure 6.12b shows the relationship between the relative velocities: V_1, the velocity of the plug relative to the barrel; and V_2, its downstream velocity relative to the screw. Hence

$$V_2 = \frac{V \tan \alpha \sec \theta_b}{(\tan \alpha + \tan \theta_b)} \tag{6.75}$$

where V is given by equation 6.5, and θ_b is the helix angle of the screw measured at the barrel surface. Now, the volumetric flow rate per screw channel, Q, is the product of V_2 and the cross-sectional area of the channel. Since the channel is relatively deep, this area is best obtained from the channel area in the plane normal to the screw axis, allowing for the presence of the flights

$$A = \frac{W}{m(W + e)} \frac{\pi}{4} [D^2 - (D - 2H)^2] \tag{6.76}$$

Projecting A through the complement of the barrel surface helix angle, the following

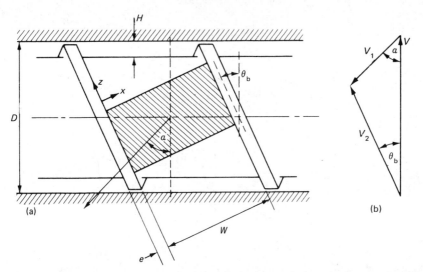

Figure 6.12 *Solid plug in the flow channel: (a) direction of motion; (b) relative velocity diagram*

expression for dimensionless flow rate is obtained

$$\pi_Q = \frac{Q}{WHV_z} = \frac{(D-H)}{D} \frac{\tan \alpha \sec^2 \theta_b}{(\tan \alpha + \tan \theta_b)} \tag{6.77}$$

In deriving this result, the screw clearance is assumed to be negligible compared with the barrel diameter in equation 6.3 for W.

It should be noted that the dimensionless flow rate is largely independent of screw size and design, unless the helix angle is changed, which is rare in practice. However, π_Q does depend on feed angle. If the polymer rotates with the screw, α is zero and there is no output. Flow rate increases with increasing α, the practical maximum (though not the mathematical maximum according to equation 6.77) occurring when $\alpha = \pi/2 - \theta_b$ and the direction of motion is normal to the screw flights. For this angle to be exceeded would require frictional forces that assisted rather than opposed the motion. If Q is known, α may be obtained from equation 6.77 for use in the equations of motion that follow.

It is convenient to define a pressure in the solid plug, p, as the direct compressive stress acting in the downstream direction. In a simple one-dimensional plug flow analysis, this pressure is assumed to be constant both across the width of the channel and through its depth. The compressive stresses at the screw and barrel surfaces are not necessarily equal to p, but may be expressed as $k_1 p$ acting on the barrel surface, $k_2 p$ on the sides of the channel and $k_3 p$ on the screw root. Figure 6.13 shows a simplified form of the system of forces acting on a portion of solid plug of length δz in the z direction, previously illustrated in figure 6.12a. The simplification is to ignore the variation of the helix angle, θ, with channel depth. In the figure, F is the net transverse force at the flights necessary to maintain equilibrium, while μ_s, μ_f and μ_b are the coefficients of friction at the screw root,

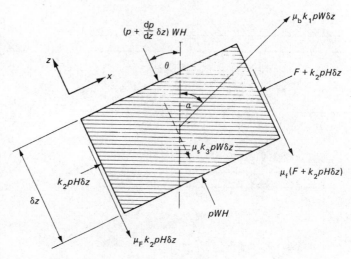

Figure 6.13   *Forces acting on the solid plug*

flight sides and barrel surfaces, respectively, and are independent of pressure and rubbing speed.

For equilibrium of forces in the z and x directions

$$\mu_b k_1 pW\,\delta z\cos(\alpha+\theta)=\mu_f F + 2\mu_f k_2 pH\,\delta z + \mu_s k_3 pW\,\delta z + \frac{dp}{dz}\,\delta z WH$$

$$\mu_b k_1 pW\,\delta z\sin(\alpha+\theta)=F$$

Elimination of the unknown F and rearrangement gives

$$\cos\alpha = K\sin\alpha + \frac{\mu_f}{\mu_b}\frac{k_2}{k_1}\frac{2H}{W}\cos\theta\,(K\tan\theta+1)$$

$$+\frac{\mu_s}{\mu_b}\frac{k_3}{k_1}\cos\theta\,(K\tan\theta+1)+\frac{H\cos\theta}{\mu_b k_1}(K\tan\theta+1)\frac{1}{p}\frac{dp}{dz}\qquad(6.78)$$

where

$$K=\frac{\tan\theta+\mu_f}{1-\mu_f\tan\theta}$$

Now, if the simplification of assuming constant θ is not made, a much more lengthy analysis leads to the following very similar form of equation

$$\cos\alpha = K\sin\alpha + \frac{\mu_f}{\mu_b}\frac{k_2}{k_1}\frac{2HE}{W}\cos\bar\theta\,(K\tan\bar\theta+E)$$

$$+\frac{\mu_s}{\mu_b}\frac{k_3}{k_1}C\cos\theta_s\,(K\tan\theta_s+C)+\frac{HE\cos\bar\theta}{\mu_b k_1}(K\tan\bar\theta+E)\frac{1}{p}\frac{dp}{dz}$$

$$(6.79)$$

where

$$K = \frac{E(\tan \bar{\theta} + \mu_f)}{(1 - \mu_f \tan \bar{\theta})}, \qquad E = \frac{D - H}{D}, \qquad C = \frac{D - 2H}{D}$$

and θ_s and $\bar{\theta}$ are the flight helix angles at the screw surface and mean channel depth, respectively. The downstream coordinate, z, is now measured at the mean channel depth, and is therefore related to the axial position by equation 6.4 with $\theta = \bar{\theta}$.

An important feature of the above analysis is the use of pressure ratios k_1, k_2 and k_3 based on the experimental observation that, when a granular material is compressed in a tube, the radial stress generated is substantially less than the applied axial stress. Tests of this type show that, for typical polymeric materials, the ratio between radial and axial stress is essentially independent of the magnitudes of the stresses, and lies in the range 0.4–0.6. There is some doubt, however, over the applicability of such data measured under relatively simple states of stress to the much more complex situation existing in plug flow in an extruder screw channel.

Both equations 6.78 and 6.79 can be expressed in the form

$$\cos \alpha = K \sin \alpha + M \qquad (6.80)$$

from which the feed angle and hence the flow rate can be found, provided the pressure and its gradient are known. Alternatively, they can be expressed in the form of a differential equation for the pressure

$$\frac{1}{p} \frac{dp}{dz} = \lambda \qquad (6.81)$$

where λ is independent of p and z. Hence

$$p = p_0 \exp(\lambda z) \qquad (6.82)$$

where p_0 is the initial pressure at $z = 0$. The need to specify an initial pressure is a difficulty not readily overcome in a simple plug-flow analysis whose assumptions imply a zero value for p_0.

Some general conclusions may be drawn from equations 6.77 and 6.79. The conveying rate is proportional to screw speed, provided the back pressure and friction coefficients are independent of speed. In order to increase the solids conveying capacity of an extruder, the ratios μ_s/μ_b and μ_f/μ_b should be decreased; that is, the screw should be smooth and the barrel rough. The conveying capacity also varies with helix angle. At very small helix angles, channel width W, and hence Q according to equation 6.77, tend to zero. As helix angles become large, both K and M in equation 6.80 increase and the feed angle, and hence Q, again tend to zero. Therefore, for any given set of material properties and pressure conditions, there is an optimum helix angle. In practice, this optimum is often close to the commonly used value of $\theta_b = 17.66°$ for a screw with a lead equal to its diameter. The simplest way to increase conveying capacity, however, is to increase the screw-channel depth. If equation 6.79 is applied to a typical screw feed section with an axial length several times its diameter, then for practical values of back pressure at the end of the section the predicted effect of back pressure on feed angle and hence conveying rate is small. Reduction of the distance over which the coulomb friction mechanism applies — for example, by melting of material in contact with the hot barrel —

increases the sensitivity of this rate to changes in back pressure. One of the main difficulties in applying any analysis of solids conveying in an extruder lies in assigning appropriate numerical values to the coefficients of friction. Some of the practical problems involved were discussed in section 3.6.

In order to illustrate the use of the above plug-flow analysis, the practical example introduced in section 6.1.1 can be extended to consider the feed section of the screw. The relevant data from this example are

$$D = 120 \text{ mm}, \ \theta_b = 17.66°, \ W = 102 \text{ mm}$$

$$V_z = 599 \text{ mm/s}, \ \dot{m} = 510 \text{ kg/h}$$

The screw channel depth in the feed section is $H = 18$ mm (implying a typical compression ratio of three based on channel depths) and the material bulk density is 1050 kg/m^3. Other material properties are

$$k_1 = k_2 = k_3 = 0.5, \ \mu_b = 0.3, \ \mu_f = \mu_s = 0.2$$

The helix angles at the mean channel depth and screw surface are

$$\bar{\theta} = \tan^{-1} \left[\frac{1}{\pi} \frac{D}{(D-H)} \right] = 20.53°$$

$$\theta_s = \tan^{-1} \left[\frac{1}{\pi} \frac{D}{(D-2H)} \right] = 24.45°$$

and the volumetric flow rate along the single screw channel is

$$Q = \frac{510}{1050} = 0.486 \text{ m}^3/\text{h} = 1.35 \times 10^{-4} \text{ m}^3/\text{s}$$

giving a dimensionless flow rate of

$$\pi_Q = \frac{Q}{WHV_z} = \frac{1.35 \times 10^{-4}}{102 \times 10^{-3} \times 18 \times 10^{-3} \times 599 \times 10^{-3}} = 0.123$$

Rearrangement of equation 6.77 gives

$$\tan \alpha = \frac{\beta \tan \theta_b}{1 - \beta}, \qquad \beta = \frac{D \pi_Q \cos^2 \theta_b}{D - H}$$

from which

$$\beta = \frac{120 \times 0.123 \times 0.9529^2}{102} = 0.131$$

$$\alpha = \tan^{-1} \left(\frac{0.131 \times 0.3183}{1 - 0.131} \right) = 2.75°$$

Hence, the various terms in equation 6.79 are

$$E = \frac{D - H}{D} = 0.85, \qquad C = \frac{D - 2H}{D} = 0.7$$

$$K = \frac{E(\tan \bar{\theta} + \mu_f)}{(1 - \mu_f \tan \bar{\theta})} = \frac{0.85(0.3745 + 0.2)}{(1 - 0.2 \times 0.3745)} = 0.528$$

$$\frac{\mu_f}{\mu_b} \frac{k_2}{k_1} \frac{2HE}{W} \cos \bar{\theta} \, (K \tan \bar{\theta} + E) = 0.196$$

$$\frac{\mu_s}{\mu_b} \frac{k_3}{k_1} C \cos \theta_s \, (K \tan \theta_s + C) = 0.399$$

$$\frac{HE \cos \bar{\theta}}{\mu_b k_1} (K \tan \bar{\theta} + E) \frac{1}{p} \frac{dp}{dz} = \frac{100}{p} \frac{dp}{dz}$$

and equation 6.79 becomes

$$\cos \alpha = 0.528 \sin \alpha + 0.196 + 0.399 + \frac{100}{p} \frac{dp}{dz} \tag{6.83}$$

For the calculated feed angle of $\alpha = 2.75°$, this result can be expressed as

$$\frac{1}{p} \frac{dp}{dz} = \lambda = 3.78 \text{ m}^{-1}$$

Using equation 6.82, the pressure ratio over one turn of the screw, which is of helical length at the mean channel depth $z = D/\sin \bar{\theta} = 0.342$ m, is

$$\frac{p}{p_0} = \exp(\lambda z) = 3.64$$

and the feed section is capable of generating substantial pressures.

It is worth noting that, if instead of equation 6.79 the simplified form displayed in equilibrium equation 6.78 had been used, a negative value of λ would have been obtained, implying that the maximum practical output had been exceeded. This demonstrates the need to take into account changes in channel geometry with depth if realistic predictions of the feeding performance of extruder screws are to be made. In the present example, if there is no forced feeding of the screw, the pressure gradient cannot be negative, and the maximum feed angle of $\alpha = 30.4°$ can be obtained from equation 6.83 when the gradient is zero. This corresponds to a π_Q value of 0.607 according to equation 6.77, and a mass flow rate of 2520 kg/h. Thus, the actual flow rate of 510 kg/h is well within the feeding capacity of the screw, a condition that is true of many screws used in practice.

Various refinements may be applied to the simple one-dimensional plug flow analysis. For example, viscous drag in thin melt films can be considered in place of coulomb friction at some or all of the interfaces between the solid plug and hot metal boundaries. In practice, there may be a significant length of channel over which viscous drag exists at the barrel surface while the screw is still cool enough for coulomb friction to apply. Although gravitational forces play an insignificant role in the conveying process once plug flow is established, they are very important in initiating the plug flow mechanism. While it is tempting to assume that the initial pressure, p_0, appearing in equation 6.82 is solely due to the pressure developed by the material in the hopper, experiments have shown that the height of material there has no significant effect on feeding performance, and that the screw channel may run only partly full of material until compaction into a solid plug occurs. Gravitational forces within the screw channel appear to provide a means of initiating pressures between the solid particles and channel boundaries, thus generating frictional forces. The cyclic nature of the supply of material from the hopper to the screw channel, resulting from the motion of the screw flight past the hopper throat, causes a corresponding cyclic variation of the initial pressure build-up. Centrifugal forces may also make a contribution, particularly in relatively large extruders.

6.2.2 Two-dimensional Plug Flow Analysis

In the simple one-dimensional plug flow analysis, the pressure or direct compressive stress in the downstream direction was assumed to vary only in this direction. The relative motions of the screw, solid plug and barrel are such, however, that frictional forces generate not only a pressure gradient in the downstream direction, but also one in the transverse direction. The pressure difference across the channel width may well be significant compared to the pressure built up in the downstream direction.

Treating pressure as a function of both channel coordinates z and x in figure 6.12a, a very lengthy analysis of equilibrium of the solid plug, including averaging stresses over the channel depth, leads to a pair of differential equations of the following form

$$k_2 \frac{\partial p}{\partial x} + \frac{\partial \tau}{\partial z} = G_1 p + H_1(x, z) \tag{6.84}$$

$$\frac{\partial p}{\partial z} + \frac{\partial \tau}{\partial x} = G_2 p + H_2(x, z) \tag{6.85}$$

The positive constant k_2 is still the ratio between local direct compressive stresses in the x and z directions. Pressure, $p(x, z)$, and shear stress, $\tau = \tau_{xz} = \tau_{zx} = \tau(x, z)$, are both averaged in the radial direction. The parameters G_1, G_2, H_1 and H_2 are determined by screw geometry, coefficients of friction, stress ratios k_1 and k_3 and the feed angle. H_1 and H_2 only appear when gravitational or centrifugal body forces are included in the analysis. If such forces are neglected and the one-dimensional assumption of insignificant variations in the x direction is introduced, then equation 6.85 reduces to the form displayed by equation 6.79.

Still neglecting body forces, τ may be eliminated from equations 6.84 and 6.85 to give

$$k_2 \frac{\partial^2 p}{\partial x^2} - \frac{\partial^2 p}{\partial z^2} = G_1 \frac{\partial p}{\partial x} - G_2 \frac{\partial p}{\partial z} \tag{6.86}$$

This hyperbolic partial differential equation yields solutions of the form

$$p = p_0 \exp(\lambda_1 z) \exp(\lambda_2 x) \tag{6.87}$$

and the pressure profiles in both the downstream and transverse directions tend to be exponential in form. If body forces are retained, solutions to the full governing equations 6.86 and 6.87 can generally only be obtained numerically, and are necessary for studying in detail both pressure initiation and the cyclic pressure variations already mentioned in connection with the one-dimensional analysis.

6.2.3 Practical Implications of Solids Conveying Analysis

A number of conclusions of practical significance can be drawn from the results of analysing the solids conveying action of single screw extruders. For example, because of the exponential variation of pressure with downstream distance, very high pressures can, in principle, be generated. This is particularly true if coulomb friction conditions exist over, say, the first six or more turns of the screw. Under such circumstances, the high pressure generating capacity ensures that the rate at which material is conveyed is largely independent of the resistance offered by the later melting and melt flow stages of the extrusion process, and is directly proportional to screw speed. In practice, however, coulomb friction conditions usually only exist over the first three or four turns of the screw before the friction mechanism becomes a viscous one. The pressure generating capacity is thereby much reduced, and the conveying rate becomes more sensitive to back pressure.

These observations have led to the development of special feed sections for extruders. These take the form of efficiently cooled barrel liners with internal axial grooves of various shapes. The grooves serve to increase the effective coefficient of friction between feedstock and barrel surface, while the cooling delays the transition from coulomb to viscous friction. Using such an arrangement, very high pressures can be generated at the end of the feed section, thereby increasing the flow rate through the machine.

Another important conclusion from the solids conveying analysis, which has already been mentioned, and which is considered again in section 6.6 in connection with surging, is the cyclic nature of the initial pressure build-up. Pressure variations, occurring at the same frequency as rotation of the screw, are unavoidably associated with the initiation of plug flow. They may lead to significant instabilities in the extrusion process as a whole.

6.3 Melting in Extruders

Mention has already been made of the onset of melting in an extruder, when a thin film of melt is formed between the hot barrel surface and the compacted plug of material in the screw channel. The first traces of melt flow into any gaps between the particles forming the solid plug, before a continuous film is created. The thickness of this film increases as the plug moves along the screw channel, and similar films are in due course formed at the screw root and flight surfaces. Experimental evidence suggests that the film at the barrel surface increases in thickness to several times the size of the clearance between the screw flight and barrel before there is any observable change in the mechanism of melting. Analysis of flow in this region, which is sometimes known as the *delay zone*, can be accomplished with the aid of viscous drag boundary conditions in the analysis for solids conveying, together with a non-isothermal treatment of the solid plug, of the type to be considered in connection with the analysis of melting.

6.3.1 Observed Melting Mechanisms

Whereas analyses of solids conveying and melt flow in extruders can be developed from fundamental mechanical principles with little or no reference to observed behaviour, attempts to treat the intervening melting process are usually based on observed mechanisms. The essence of the experimental method used is first to achieve the required steady running conditions and then to stop the screw and cool the barrel as rapidly as possible. After extracting the screw from the barrel, the helical strip of solidified polymer is removed from the screw channel and sectioned. Various visual aids may be used to help distinguish between material melted in the presence of shear before the screw was stopped, and that which has melted subsequently by thermal conduction alone. These aids include the use of granules of mixed colours, the addition of dry colouring to coat individual granules and changing feedstock colour just prior to stopping the machine.

Typical sections through the contents of the screw channel are as shown schematically in figure 6.14a and b, views from a stationary screw equivalent to figure 6.3. In figure 6.14a, the screw surface temperature, T_s, is below the melting temperature, T_m, of the polymer and no melt film is present at the screw. Such a situation usually exists for a short distance before T_s exceeds T_m, and a film is formed at the screw as shown in figure 6.14b. Both sectional views are somewhat idealised, in that the channel is actually curved, and the corners of the solid bed become rounded. The term 'solid bed' customarily used in connection with melting is equivalent to the term 'solid plug' in conveying analyses.

There are up to four distinct regions in the screw channel cross-section — the upper and lower melt films at the barrel and screw surfaces, respectively, the solid

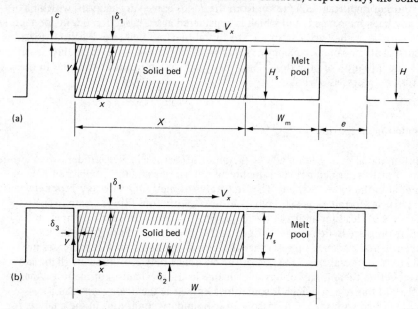

Figure 6.14   *Melting mechanisms: (a) without lower melt film (cool screw); (b) with lower film formed (hot screw)*

bed and the melt pool. The lower melt film could be regarded as two regions, the film along the screw root and the one on the trailing side of the flight. Owing to the proximity of the heated barrel and the intense shear, much of the melting occurs in the upper melt film. The motion of the barrel relative to the solid bed sweeps the melt so formed into the melt-pool region between the bed and the leading side of the flight. Clearly, this pool is only likely to be formed when the thickness of the upper melt film, δ_1 in figure 6.14, exceeds the clearance between the flight and barrel, and the flight can act as a scraper. While it is to be expected that δ_1 varies with both the transverse coordinate x and downstream coordinate z, such variations are difficult to measure. The thickness δ_1 remains small as newly melted material is transferred to the pool. The flow in this pool is a combination of clockwise circulation and downstream motion due to the dragging actions of both the barrel and bed moving relative to the screw, together with the effect of the downstream pressure gradient that normally exists.

If melt films are formed at the screw surfaces, their thicknesses, δ_2 at the screw root and δ_3 at the flight, are observed to be independent of x and y, respectively. As the motion of the solid bed relative to the screw is in the downstream direction, the lower films grow in thickness in this direction. This growth can be considerable, as a result of the retention of all melted material. Usually, the main change exhibited by successive cross sections along the screw is a reduction in the width of the solid bed, with a corresponding increase in the melt pool width. With screws having rapidly reducing channel depths in their compression sections, however, it is possible to observe a temporary increase in solid bed width as the bed is forced into the resulting wedge.

Completion of melting, which is often not achieved until well into the metering section, can occur in various ways. If the process is stable and the bed remains continuous in the downstream direction, the corners of the bed become rounded and either its width or depth diminish to negligible proportions. The final disappearance of the bed can be quite rapid once the temperature of the bulk of the solid has risen close to the melting temperature. Very often the melting process becomes unstable, in the sense that the bed suffers breaks along planes normal to the downstream direction. This phenomenon is known as solid bed break-up and is discussed in more detail in section 6.3.4. The resulting pieces of compacted solid then decrease in size, often as much by thermal conduction as by shearing in thin melt films.

Although the above mechanism of melting appears to be the most common, there are some significant exceptions. With material supplied in powder form, particularly in the case of rigid polyvinyl chloride, a form of melt pool may develop against the trailing edge of the flight. With polyvinyl chloride, the melting behaviour is also dependent on the type and quantity of lubricant used. Another variant of the usual melting mechanism has been observed in relatively large extruders, in which the melt remains in relatively thick films surrounding the solid bed.

6.3.2 The Simple Tadmor Model for Melting

Returning to the melting mechanism of the type shown in figure 6.14, further dimensions not previously defined are the melt-pool width, W_m, solid bed width,

X, and depth, H_s. The downstream velocity of the solid bed relative to the screw, V_{sz}, is its dominant velocity component, and is assumed to be independent of x and y. Many of the assumptions introduced earlier for melt flow analyses are also applicable in the derivation of melting models. Assumptions 1–15 and 17 are reasonable, particularly for melt flow regions, and assumption 16 is applicable to at least the flow in the melt pool. Following assumption 11 — constant melt specific heat, C_{pm} — it is also reasonable to assume a constant specific heat, C_{ps}, for the polymer in the solid state and that there is a sharp melting point, T_m, and an associated latent heat of fusion, λ. These thermal-property assumptions were discussed in section 3.1.

One of the simplest theoretical models for melting, based on the mechanism shown in figure 6.14a, is often referred to as the Tadmor model, after its originator. The downstream bed velocity, V_{sz}, is assumed to be independent of z, and its value is calculated as that necessary for the solid plug prior to melting to give the prescribed total mass flow rate along the channel. As a consequence of this considerable simplification, the motion of the bed is prescribed, and conditions for its equilibrium need not be examined. A further simplification is the omission of the lower melt film although, given the constant V_{sz} assumption, this merely implies that the rate of melting is determined by the upper melt film alone. In its simplest form, the Tadmor model assumes that the thickness of the upper melt film, δ_1, is independent of x and that the flow there can be treated as fully developed isothermal newtonian drag flow. It should be clear from section 5.5 that, while the omission of pressure gradient and thermal development effects may be reasonable for flow in a thin film of this type, the isothermal and newtonian assumptions may lead to significant errors.

The velocities of the barrel relative to the solid bed are $(V_z - V_{sz})$ in the downstream direction, and V_x in the transverse direction, giving a resultant relative velocity of

$$V_r = [(V_z - V_{sz})^2 + V_x^2]^{1/2} \qquad (6.88)$$

Figure 6.15 illustrates the relationship between the various velocities, including the angle β between V_r and the downstream direction. With a constant shear rate, V_r/δ_1, throughout the upper film, where μ_1 is the constant melt viscosity, energy equation 4.76 becomes

$$k_m \frac{d^2 T}{dy^2} = -\mu_1 \left(\frac{V_r}{\delta_1} \right)^2 \qquad (6.89)$$

where k_m is the melt thermal conductivity, and coordinate y is measured radially outward from the surface of the solid bed. This equation may be integrated with the boundary conditions $T = T_m$ at $y = 0$, $T = T_b$ at $y = \delta_1$, to give

$$T - T_m = \frac{\mu_1 V_r^2}{2 k_m} \frac{y}{\delta_1} \left(1 - \frac{y}{\delta_1} \right) + \frac{y}{\delta_1} (T_b - T_m) \qquad (6.90)$$

the first term on the right-hand side being due to viscous dissipation and the second to thermal conduction. The heat flow rate per unit area into the solid bed is

$$k_m \left(\frac{dT}{dy} \right)_{y=0} = \frac{\mu_1 V_r^2}{2 \delta_1} + \frac{k_m}{\delta_1} (T_b - T_m) \qquad (6.91)$$

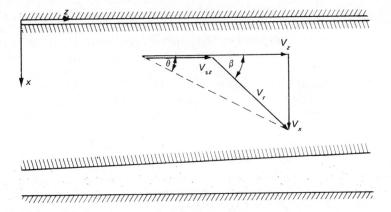

Figure 6.15   *Relative velocity diagram for the solid bed*

and is used both to melt the solid and to heat up the bulk of the solid bed. Neglecting the latter effect, the rate of melting per unit area from the top of the bed may be obtained as

$$\omega_1 = \frac{1}{\lambda \delta_1} \left[\frac{\mu_1 V_r^2}{2} + k_m (T_b - T_m) \right] = \frac{C_1}{\lambda \delta_1} \qquad (6.92)$$

where C_1 is merely a convenient shorthand for the bracketed expression. Because δ_1 remains locally independent of z, all material melted into the upper film is dragged into the melt pool. Neglecting flow over the flight

$$\omega_1 X = \tfrac{1}{2} \rho_m V_x \delta_1 \qquad (6.93)$$

where ρ_m is melt density. Combining equations 6.92 and 6.93

$$\delta_1 = \left(\frac{2C_1}{\lambda \rho_m V_x} \right)^{1/2} X^{1/2}, \qquad \omega_1 = \left(\frac{C_1 \rho_m V_x}{2\lambda} \right)^{1/2} X^{-1/2} \qquad (6.94)$$

Note that δ_1 and ω_1 are respectively proportional and inversely proportional to the square root of the solid bed width. Now the downstream mass flow rate in the bed decreases as a result of the loss of material by melting

$$\frac{d}{dz} (\rho_s V_{sz} X H_s) = -\omega_1 X \qquad (6.95)$$

where ρ_s is the bulk density of the solid bed. Making the approximation $H_s = H$, this equation becomes

$$\rho_s V_{sz} \frac{d}{dz} (XH) = - \left(\frac{C_1 \rho_m V_x}{2\lambda} \right)^{1/2} X^{1/2} = -C_2 X^{1/2} \qquad (6.96)$$

If the channel depth, H, is constant, differential equation 6.96 can be solved to give

$$\frac{X}{W} = \left(1 - \frac{C_2 z}{2 \rho_s V_{sz} H W^{1/2}} \right)^2 \qquad (6.97)$$

where melting starts at $z = 0$ with $X = W$, the channel width. This reduced solid bed width, X/W, is a convenient parameter for describing the degree of melting achieved at a particular downstream position. Equation 6.96 can also be solved analytically when H is a linear function of z, which it is in the compression section of a screw.

While the above model is too simple to give reliable quantitative results, it does show qualitatively the type of melting performance to be expected in practice. For example, from equation 6.97 the helical length of screw channel required for melting to be completed is

$$z_m = \frac{2\rho_s V_{sz} H W^{1/2}}{C_2} = 2\rho_s V_{sz} H \left(\frac{2\lambda W}{C_1 \rho_m V_x} \right)^{1/2} \tag{6.98}$$

As V_{sz} is directly proportional to the mass flow rate of polymer through the extruder, it follows that the melting length is also proportional to this output rate. Thus, any process changes that improve output for a given screw speed tend also to reduce extrudate quality, because the length of melt filled channel in which mixing can occur is reduced. Note also that equation 6.97 predicts X to be a parabolic function of z for a channel of constant depth.

6.3.3 An Improved Melting Model

Many improvements can be made to the simple Tadmor model. These include improved analyses for flow and heat transfer in the melt films and solid bed. The most important refinement, however, is to relax the assumption of constant downstream solid bed velocity to provide a more realistic description of solid bed motion and deformation. Clearly, the bed suffers substantial deformation in order that its width may decrease while most of the melting is taking place at its upper surface. This deformation need not be confined to the x, y plane: elongation of the bed in the z direction may contribute to its reduction in width. V_{sz} should be regarded as a slowly varying function of z. Consequently, equilibrium of the bed must be considered, and the lower melt films at the screw surfaces become important in terms of the shear stresses exerted on the bed, in addition to the melting caused. The following model has been shown to offer a good compromise between accurate prediction of actual melting behaviour and the cost of computing the numerical solutions involved.

With variable downstream bed velocity, mass balances must be established for the various regions of the flow. For the upper melt film

$$\begin{pmatrix} \text{Rate of change} \\ \text{of downstream} \\ \text{mass flow rate} \end{pmatrix} = \begin{pmatrix} \text{Rate of melting} \\ \text{over interface} \\ \text{with solid bed} \end{pmatrix} - \begin{pmatrix} \text{Net transverse} \\ \text{flow rate out} \\ \text{of melt film} \end{pmatrix}$$

$$\frac{d}{dz}(m_{1z}X) = \omega_1 X - (m_{1x} - m_{fx}) \tag{6.99}$$

where m_{1z} and m_{1x} are the mass flow rates in the upper film in the z and x directions, respectively, per unit width of film in the x and z directions, respectively. Assuming drag flow and a linear velocity profile in the clearance as discussed in section 6.1.5, the leakage mass flow rate over a flight in the transverse direction per unit downstream distance is

$$m_{fx} = \tfrac{1}{2}\rho_m V_x c \qquad (6.100)$$

Equation 6.99 is a generalised form of equation 6.93, which allows for flow over the flights, downstream development of the upper melt film thickness and the possibility of nonlinear velocity profiles in this film. It does, however, involve the assumption that the melting rate, ω_1, and by implication the film thickness, δ_1, are independent of the position x across the width of the bed.

In the cases of the lower melt films, which are each assumed to be of uniform thickness at a particular channel cross section, all the melt is retained, and the mass balance conditions are

$$\frac{d}{dz}(m_{2z}X) = \omega_2 X, \qquad \frac{d}{dz}(m_{3z}H_s) = \omega_3 H_s \qquad (6.101)$$

m_{2z} and m_{3z} are the downstream mass flow rates per unit width of film in the screw root and flight films, respectively, ω_2 and ω_3 being the corresponding melting rates per unit area. Neglecting melting at the interface between the bed and melt pool, equation 6.95 for the solid bed mass balance is now

$$\frac{d}{dz}(\rho_s V_{sz} X H_s) = -\omega_1 X - \omega_2 X - \omega_3 H_s \qquad (6.102)$$

where

$$H_s = H - \delta_1 - \delta_2$$

The downstream mass flow rate in the melt pool, M_{mz}, depends on its size, the velocities of the barrel and solid bed relative to the screw and the downstream pressure gradient, P_z. An over-all mass balance is used to equate the total flow rate along a screw channel, M_T, which is assumed to be a known quantity and, for steady flow, to be independent of both time and z, to the sum of the flow rates in the melt films, solid bed and melt pool

$$M_T = m_{1z}X + m_{2z}X + m_{3z}H_s + \rho_s V_{sz} X H_s + M_{mz} - \tfrac{1}{2}\rho_m W V_z c \qquad (6.103)$$

The argument for including the last leakage flow term was developed in section 6.1.5, and its form is derived from equation 6.74.

In addition to the mass balances, a balance between the forces due to the downstream pressure gradient acting on the cross-section of the bed and the shear forces acting on its surfaces gives

$$P_z H_s X = \tau_{1z}X - \tau_{2z}X - \tau_{3z}H_s \qquad (6.104)$$

where τ_{1z}, τ_{2z} and τ_{3z} are the shear stresses in the z direction at the bed surfaces in the upper and lower melt films. The much smaller stresses on the bed due to the melt pool are neglected. Note that while τ_{1z} due to the dragging motion of the barrel relative to the bed is in the positive downstream direction, its effect is opposed by the shear forces in the lower melt films. Indeed, for the usual case of a positive pressure gradient during melting, the pressure force on the bed cross-section also opposes the shear force in the upper film. Clearly, V_{sz} can never exceed V_z or the sign of τ_{1z} would be reversed, and in practice the ratio V_{sz}/V_z is relatively small.

In order to determine parameters m_{1z}, m_{1x}, m_{2z}, m_{3z}, ω_1, ω_2, ω_3, τ_{1z}, τ_{2z}, τ_{3z} and M_{mz} in governing equations 6.19 and 6.101–6.104, detailed analyses of flow and heat transfer in the films, bed and melt-pool regions must be performed. In the case of the solid bed, no attempt is made to analyse internal deformations. The temperature profile within the bed may be obtained with the aid of a simplified form of energy equation 4.17, which is analogous to equation 6.61 used to describe developing melt flow

$$\rho_s C_{ps} V_{sz} \frac{\partial T}{\partial z} = k_s \frac{\partial^2 T}{\partial y^2} \qquad (6.105)$$

where k_s is the thermal conductivity of the solid bed. Note that thermal conduction in the transverse and downstream directions is ignored. The boundary conditions are $T = T_m$ at the upper and lower faces of the bed, together with an initial condition of ambient temperature throughout at the beginning of melting. The solid bed temperature profile can be obtained by a numerical finite difference treatment of equation 6.105 very similar to that described in appendix B for thermally developing melt flow, but with the simplification of having no dissipation term. As a result of neglecting bed temperature variations in the x direction, it is not possible to analyse realistically the melting into the small film at the screw flight. It is therefore convenient to assume $\delta_3 = \delta_2$, $m_{3z} = m_{2z}$, $\omega_3 = \omega_2$ and $\tau_{3z} = \tau_{2z}$. The side film can be treated as an extension of the one at the screw root, and equations 6.101 become

$$\frac{d}{dz}[m_{2z}(X + H_s)] = \omega_2(X + H_s) \qquad (6.106)$$

A thorough analysis of flow and heat transfer in the melt pool presents problems at least as great as those already discussed for flow in a melt filled channel. In addition to a net influx of melt and two moving boundaries instead of one, the assumption of an infinitely wide channel (assumption 20) is much less justifiable. In order to avoid an excessive expenditure of computational effort on the melt-pool analysis, the rather crude assumption of isothermal newtonian flow is made. Using equations 6.45–6.47, the volumetric flow rates in the pool due to the various drag and pressure flow effects can be expressed in terms of shape factors F_D and F_P that are functions only of the flow channel depth-to-width ratio. For example, as a result of the motion of the barrel relative to the screw

$$Q_1 = \tfrac{1}{2} W_m H V_z F_D\left(\frac{H}{W_m}\right) \qquad (6.107)$$

and as a result of the motion of the solid bed

$$Q_2 = \tfrac{1}{2} H W_m V_{sz} F_D\left(\frac{W_m}{H}\right) \qquad (6.108)$$

Also, as a result of the downstream pressure gradient

$$Q_3 = -\frac{W_m H^3 P_z}{12\bar{\mu}} F_P\left(\frac{H}{W_m}\right) = -\frac{H W_m^3 P_z}{12\bar{\mu}} F_P\left(\frac{W_m}{H}\right) \qquad (6.109)$$

where $\bar{\mu}$ is a suitable mean viscosity, evaluated at the mean shear rate in the pool

$$\bar{\gamma} = \left[\left(\frac{V_z}{H} \right)^2 + \left(\frac{V_{sz}}{W_m} \right)^2 \right]^{1/2} \qquad (6.110)$$

and the bulk mean temperature of the melt entering from the upper film. Note the two alternative forms of the result in equation 6.109 according to whether H and W_m are taken as the channel depth and width, or vice versa. Finally, the downstream mass flow rate in the melt pool may be found by superimposing the three components

$$M_{mz} = \rho_m (Q_1 + Q_2 + Q_3) \qquad (6.111)$$

The analysis of melt flow in an intensely sheared thin film was considered in section 5.5, in preparation for the present melting analysis, and a largely analytical solution for nonisothermal, non-newtonian drag flow was described in section 5.5.1. Consider first the flow in the lower melt film of local thickness δ_2, which is assumed to be a drag flow associated with the downstream velocity component, V_{sz}, of the barrel relative to the screw. The local downstream mass flow rate per unit width is

$$m_{2z} = \rho_m V_{sz} \delta_2 (1 - \pi_{Q2}) \qquad (6.112)$$

where π_{Q2} is the dimensionless volumetric flow rate obtained from equation 5.112 for the conditions prevailing in the lower film. Note that $(1 - \pi_{Q2})$ is used rather than π_{Q2} because the analytical solution relates to a film in which the melting interface is fixed and the other boundary moves, whereas in the present case the reverse is true. There is some difficulty in defining the film boundary temperature at the screw surface. Although in the metering section of an extruder screw the screw temperature is approximately equal to the barrel temperature as discussed in section 6.1.3, at the feed end the screw temperature is usually much lower than that of the heated barrel. Practical experience shows that an empirical function can be used to describe the variation of screw temperature with downstream position, which allows this temperature to rise exponentially to that of the barrel. In addition to mass flow rate, both the temperature gradient at the melting interface and the shear stress in the lower film are required

$$\left(\frac{dT}{dy} \right)_{y=0} = \frac{1}{b\delta_2} \left(\frac{dT^*}{dY} \right)_{Y=0}, \qquad \tau_{2z} = \bar{\tau}_2 S \qquad (6.113)$$

where S and the dimensionless temperature gradient are obtained from the analytical solution, and y is in the outward normal direction to the solid-bed surface, which is the inward radial direction of the screw.

In the upper melt film of local thickness δ_1, the relative velocity between the barrel and solid bed is V_r at an angle β to the downstream direction, as shown in figure 6.15. As freshly melted material is entrained in the direction of V_r, there must be a corresponding increase in film thickness. This is an unavoidable consequence of the assumption of drag flow in the upper film, the justification fo which was provided in section 5.5. As the magnitude of V_{sz} rarely exceeds half that of V_z, V_r is not predominantly in either the downstream or transverse directions. While δ_1 should be treated as a function of both x and z, this is likely to prove very costly. While δ_1 must vary across the width of the bed, the variation is assumed to be sufficiently small for an average value to be used, and the present model treats δ_1 as a function of z only. An alternative approach that has been used is to treat δ_1 as a function of x only, but the computational problems are more severe and the results in general not significantly superior. The local mass flow rate per unit width, melting-interface temperature gradient and shear stress in the upper film may be obtained from the dimensionless results of the analytical solution as

$$m_1 = \rho_m V_r \delta_1 \pi_{Q1}, \quad \left(\frac{dT}{dy}\right)_{y=0} = \frac{1}{b\delta_1} \left(\frac{dT^*}{dY}\right)_{Y=0}, \tau_1 = \bar{\tau}_1 S \qquad (6.114)$$

where both m_1 and τ_1 are in the direction parallel to V_r. Hence, the required mass flow rate and shear stress components may be obtained as

$$m_{1x} = m_1 \sin \beta \qquad (6.115)$$

$$m_{1z} = m_1 \cos \beta + \rho_m \delta_1 V_{sz} \qquad (6.116)$$

$$\tau_{1z} = \tau_1 \cos \beta \qquad (6.117)$$

Finally, the melting rates at the interfaces between the solid bed and melt films may be determined from the known temperature gradients. For the upper melt film

$$\omega_1 \lambda_1 = k_m \left(\frac{\partial T}{\partial y}\right)_m - k_s \left(\frac{\partial T}{\partial y}\right)_s \qquad (6.118)$$

where the temperature gradients are at the melting interface, for the melt and solid, respectively, the latter being obtained from the finite difference solution for the bed temperature profile. Similarly, for the lower film

$$\omega_2 \lambda_2 = k_m \left(\frac{\partial T}{\partial y}\right)_m - k_s \left(\frac{\partial T}{\partial y}\right)_s \qquad (6.119)$$

where y is still the local outward normal to the surface of the solid bed. Heating up of the solid as it moves towards the melting interface, and heating of the freshly melted material to the mean temperatures of the films — both effectively forms of convection in the y direction — can be accounted for by defining the latent heats of fusion, λ_1 and λ_2, not as the value λ but as

$$\lambda_1 = \lambda + C_{ps}(T_m - \bar{T}_{sb}) + C_{pm}(\bar{T}_1 - T_m) \qquad (6.120)$$

$$\lambda_2 = \lambda + C_{ps}(T_m - \bar{T}_{sb}) + C_{pm}(\bar{T}_2 - T_m) \qquad (6.121)$$

where $\bar{T}_{sb}$ is the bulk mean temperature of the solid bed, and $\bar{T}_1$ and $\bar{T}_2$ are the bulk mean melt temperatures in the upper and lower films, obtained from the analytical solution. For a polymer that does not exhibit a sharp melting point — which, as discussed in section 3.1, is relatively common — equations 6.120 and 6.121 can be interpreted as the over-all changes in specific enthalpy between the mean temperature of the solid and those of the melt films.

A computational procedure for solving the melting model equations may be outlined as follows.

(1) Given the current values of δ_1, δ_2, V_{sz}, X and P_z, and also the temperature profile in the solid bed at a particular channel cross section, obtain the mass flow

rates m_{1x}, m_{1z} and m_{2z} from equations 6.115, 6.116 and 6.112, also the melting rates ω_1 and ω_2 from equations 6.118 and 6.119, and the shear stresses τ_{1z} and τ_{2z} from equations 6.117 and 6.113.

(2) Calculate the solid bed temperature profile at the next channel cross section, one small finite difference step downstream, by the type of method outlined in appendix B.

(3) In order to determine the values of δ_1, δ_2, V_{sz}, X and P_z at the new cross section, use differential equations 6.99, 6.102 and 6.106, together with mass and force balance equations 6.103 and 6.104, to step downstream by the method outlined in appendix C.

(4) Compute the pressure at the new position using the mean value of its gradient over the downstream step.

(5) Repeat items 1–4 for successive downstream positions until either melting is complete, the end of the screw is reached or the calculation procedure breaks down.

Melting is terminated, often well into the metering section, when either X decreases to zero, or δ_2 and to some extent δ_1 grow to make H_s zero. Sometimes the solid bed velocity increases so rapidly towards the end of melting that the calculation scheme becomes unstable. This behaviour, which appears to correspond to a real physical phenomenon, is discussed in the next subsection. The size of the downstream steps should be kept small to reduce the risk of instability, about 100–200 over the length of a typical screw being appropriate.

The initial conditions required to start the solution procedure may be difficult to define. Owing to the cooling usually applied to the feed pocket, the temperatures of both the screw and barrel surfaces are below the melting point of the polymer, and a solids-conveying analysis is appropriate. Normally after about three or four turns of the screw, the barrel temperature has increased sufficiently for melting to start. Formation of the melt film on an uncooled screw usually occurs one or two turns later. At the first channel cross section to be included in the analysis, initial values are required for V_{sz} and X. Assuming that the solid bed is the only flow region present, V_{sz} is found from over-all mass balance equation 6.103, and X is equal to the channel width.

The main deficiency of the present melting model is its inability to make reliable predictions of pressure; this is apparently a consequence of the rather crude model used to describe flow in the melt pool. Other predictions also become unrealistic towards the end of melting, where the flow is dominated by the melt pool.

The following practical example illustrates the use of the model and compares the results obtained with experimental measurements. A polystyrene supplied in granular form has the following properties

$$\rho_s = 1050 \text{ kg/m}^3, C_{ps} = 2010 \text{ J/kg °C}, k_s = 0.20 \text{ W/m °C}$$

$$\lambda = 25.9 \text{ kJ/kg at } T_m = 140 \text{ °C}$$

$$\rho_m = 990 \text{ kg/m}^3, C_{pm} = 2000 \text{ J/kg °C}, k_m = 0.21 \text{ W/m °C}$$

$$\mu_0 = 10.8 \text{ kN s/m}^2 \text{ at } \gamma_0 = 1 \text{ s}^{-1} \text{ and } T_0 = 200 \text{ °C}$$

$$n = 0.36, b = 0.022 \text{ °C}^{-1}$$

This material is processed by an extruder of diameter 63.5 mm. The screw is single start with a flight pitch equal to its diameter, and the axial lengths of the feed, compression and metering sections are 317 mm, 635 mm and 635 mm, respectively. Channel depth varies from 8.38 mm in the feed section to 2.79 mm in the metering section, the flight width is $e = 6.35$ mm and the radial clearance $c = 0.10$ mm. The operating conditions involve a constant barrel temperature of $T_b = 200\ ^{\circ}C$, and screw speeds of either $N = 60$ or 100 rev/min with total mass flow rates of $M_T = 47.2$ kg/h or 77.7 kg/h, respectively.

Figures 6.16 and 6.17 show reduced solid bed width, X/W, and relative bed velocity, V_{sz}/V_z, plotted as functions of downstream position for the two sets of operating conditions. In each case, both measured and predicted values are shown. Solid bed widths and velocities are measured after screw extraction, the latter being deduced from the directions of surface striations visible on the solidified upper melt film, a technique that is most effective with polystyrene. Other variables, such as film thicknesses, are difficult to measure accurately.

Taking reduced solid bed widths first, the agreement between experiment and the theoretical model is excellent in the early stages of melting. Note that the model predicts a variation similar to the parabolic form of equation 6.97, derived from the Tadmor model for a channel of constant depth. In the later stages of melting, agreement between theory and experiment deteriorates. While at the lower screw speed the width of the bed starts to fluctuate, at the higher speed the bed breaks along planes normal to the downstream direction. Hence, the plotted zero measured values of X/W. Clearly, the melting model cannot be expected to apply under such conditions.

Turning to the relative solid bed velocities, the agreement between experiment and the theoretical model is again good, although the measured velocities are consistently higher than the predicted ones. This discrepancy can be eliminated without altering the shapes of the curves by a different choice of the effective solid

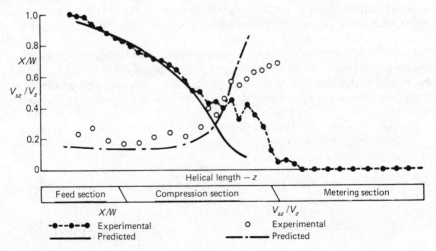

Figure 6.16   *Variations of reduced solid bed width and relative bed velocity for a screw speed of N = 60 rev/min*

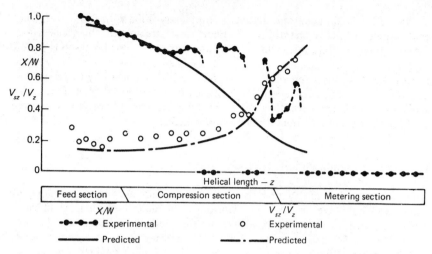

Figure 6.17   *Variations of reduced solid bed width and relative bed velocity for a screw speed of N = 100 rev/min*

bed bulk density. Note that initially V_{sz}/V_z is of the order of 0.2 for both screw speeds, implying a value for the angle β in figure 6.15 of about 22°, which means that downstream development of the upper melt film is more significant than its transverse development. After remaining nearly constant for some distance, the solid bed velocity starts to increase, the region of substantial acceleration corresponding to the onset of bed break-up. Experience of a much wider range of melting experiments and model predictions indicates both that it is unusual for the solid bed to remain continuous as in figure 6.16, and that solid bed break-up is almost invariable associated with significant bed acceleration. The present model therefore provides a valuable means of studying extruder melting performance and stability.

6.3.4 Solid Bed Break-up

The frequently observed phenomenon of periodic solid bed break-up towards the end of melting presumably occurs because the state of stress within the bed departs too far from a state of isotropic compression, although in the absence of a detailed stress analysis for the bed no formal criterion can be established. As indicated by equation 6.104, the equilibrium of the bed as a whole involves a balance between forces associated with the downstream pressure gradient acting on the cross section of the bed, and the shear forces acting on its sides. As melting proceeds, the upper melt film remains relatively thin while the lower films increase in thickness, and unless the pressure gradient increases significantly, the bed velocity must increase to help balance the shear forces in the films. An increase in the pressure gradient, particularly towards the end of melting, where the melt pool is relatively wide, would reduce the mass flow rate in the pool and require an increase in the bed velocity to maintain the over-all mass balance.

Once a break has occurred, the gap is completely filled with melt and the new end of the continuous part of the bed moves downstream until the accelerating forces applied to it are sufficient to cause another break. Bed break-up is therefore a periodic instability occurring at frequencies that are of the order of five to ten times lower than the screw rotation frequency. One practical way to prevent bed break-up is to apply screw cooling, which delays formation of a melt film at the screw surface, thereby maintaining a more stable balance between the shear forces on the bed but reducing the output from the extruder.

6.4 Power Consumption in Extruders

An important result of any analysis of extruder performance is an estimate of the machine's total power requirement. This power includes both the mechanical power supplied through the screw drive and the thermal power supplied by the barrel heaters. The drive power is normally the greater of the two, and must be matched both to the capacity of the motor and to the torsional strength of the screw. Heater capacities are usually determined more by the need for rapid heating of the barrel before the machine can be used, rather than the heat required in normal running. Indeed, a net cooling of the barrel may be necessary, particularly on large machines.

The mechanical power consumption can be calculated as the rate of working at the interface between the material in the screw channel and the barrel. This rate is

$$\dot{E} = V_z \tau_{zy}{}' + V_x \tau_{xy}{}' \tag{6.122}$$

per unit area, where the shear stresses are evaluated at the barrel. The total power is found by integrating this expression over the entire barrel surface, not only for the solids conveying, melting and melt flow regions, but also over the leakage flow in the clearance. The values of the shear stresses used for this purpose clearly depend on the particular theoretical model used.

For illustration purposes, consider the isothermal newtonian melt flow model derived in section 6.1.2. From equations 6.36 and 6.42

$$\tau_{zy}{}' = \mu \left(\frac{dw}{dy} \right)_{y=H} = \mu \frac{V_z}{H} + \frac{HP_z}{2} \tag{6.123}$$

$$\tau_{xy}{}' = \mu \left(\frac{du}{dy} \right)_{y=H} = \frac{4\mu V_x}{H} \tag{6.124}$$

The power consumption per unit channel length in the downstream direction is

$$E_z = \int_0^W \dot{E} \, dx = W \left(\frac{\mu V_z{}^2}{H} + \frac{V_z H P_z}{2} + \frac{4\mu V_x{}^2}{H} \right) \tag{6.125}$$

Introducing the mean shear stress, $\bar{\tau}$, defined at the mean shear rate, $\bar{\gamma} = V_z/H$, a dimensionless channel power consumption may be defined as

$$\pi_E = \frac{E_z}{WV_z\bar{\tau}} = 1 + \frac{\pi_P}{2} + 4 \tan^2 \theta \tag{6.126}$$

and is the ratio between the actual power consumption and that which would occur in pure drag flow. The three components of this expression for π_E can be identified as due to downstream drag flow, downstream pressure flow and transverse flow, respectively. The recirculating transverse flow clearly absorbs a significant proportion of the total power, often as least as much as the downstream pressure flow, the typical magnitude of π_P being less than unity. Section 6.7.4 provides an example of power consumptions computed using the more realistic developing melt flow model described in section 6.1.4.

Turning to the leakage flow, if isothermal drag flow conditions are assumed, equation 6.122 may be integrated across the flight width to find the power consumed per unit downstream length of flight as

$$E_z' = \frac{\mu' V^2 e}{c} \tag{6.127}$$

where μ' is the clearance viscosity, evaluated at the barrel temperature and mean shear rate, V/c. Typical results obtained using this equation suggest that a substantial proportion of the total drive power is dissipated in the clearance. As indicated in section 6.1.5, however, there is evidence to suggest that this is not necessarily so in practice, because of the onset of slip and the absence of melt in the clearance. More realistic predictions of power requirements are often obtained by ignoring leakage flow.

6.5 Mixing in Extruders

Mixing in melt flow processes in general was discussed in section 4.7. The method derived for calculating bulk mean distributive mixing may now be applied to flow in an extruder screw channel. Adapting equation 4.84, the bulk mean mixing per unit downstream length of channel is

$$\frac{d\overline{M}}{dz} = \frac{1}{Q} \int_0^H \int_0^W (4I_2)^{1/2} \, dx \, dy \tag{6.128}$$

where Q is the volumetric flow rate. The distribution of the second invariant of the rate-of-deformation tensor, I_2, depends on the particular theoretical model used to describe the melt flow. Now, the main purpose of evaluating mixing is to be able to compare the performances of different machines. Because most extruder screws have the same helix angle, as a first approximation it is possible to ignore the contribution of transverse flow in the screw channel and evaluate mixing for downstream flow only.

If transverse flow is ignored, and the melt flow assumed to be one dimensional, an estimate of the bulk mean mixing is given by

$$\frac{d\overline{M}}{dz} = \frac{W}{Q} \int_0^H \left| \frac{dw}{dy} \right| dy = \frac{W V_z}{Q} = \frac{1}{H \pi_Q} \tag{6.129}$$

provided the velocity gradient does not change sign. Thus, for a downstream length,

Z, of constant depth channel, the total bulk mean mixing can be obtained as

$$\bar{M} = \frac{Z}{H\pi_Q} \qquad (6.130)$$

Note that $\bar{M}$ is proportional to screw length, and inversely proportional to both channel depth and dimensionless flow rate. This result indicates the changes in machine design or operating conditions necessary to improve mixing performance.

Consider a typical extruder with a metering-section length ten times its diameter, a helix angle of $\theta = 17.66°$ and a channel depth-to-diameter ratio of $H/D = 0.05$, processing melt at a dimensionless rate $\pi_Q = 0.4$. From equation 6.4, the helical length of the metering section is

$$Z = \frac{10D}{\sin\theta} = 33.0D$$

and the mixing imparted by this section is

$$\bar{M} = \frac{33.0D}{0.05D} \times \frac{1}{0.4} = 1650$$

The large size of this number, which is the mean shear strain applied to the melt, helps to explain why screw extruders are efficient mixers.

In order to produce further improvement in the mixing performance of single-screw extruders, numerous types of special mixing devices have been added to conventional screws. These include smear heads, blisters, turbine mixing heads, reverse flighted sections and channel dams. The functions of such devices appear to be threefold: to apply some distributive mixing, to apply dispersive mixing and to disturb established flow and temperature profiles.

6.6 Surging in Extruders

So far, only steady extrusion has been considered. In practice, however, surging that involves periodic fluctuation of flow rate and pressures is sometimes a serious problem. A thorough analysis of surging requires the retention of the time-dependent terms in the equations governing flow. Also, the viscoelastic nature of polymers is more important in unsteady flow. Nevertheless, it is possible to establish some of the apparent causes of surging.

Considering the feed end of extruders, as indicated in section 6.2.1, pressure fluctuations are unavoidably associated with the solids conveying process. These fluctuations occur at the same frequency as the screw rotation, but may not be significant in magnitude compared to the over-all pressures generated. Instabilities in the melting process, associated with periodic break-up of the solid bed, were discussed in sections 6.3.3 and 6.3.4, and occur at frequencies that are of the order of five to ten times lower than the screw rotation frequency. Finally, very slow fluctuations over periods of minutes or even hours may be caused by instabilities in the temperature control systems or changes in the environment.

6.7 Over-all Performance and Design of Extruders

Having considered various features of the behaviour of single-screw extruders, including solids conveying, melting, melt flow, mixing and surging, it now remains to draw these together in considering over-all machine performance and design.

6.7.1 Over-all Extruder Performance

If analyses of solids conveying, melting and melt flow are to be useful in studying extruder performance, they must be combined in order to predict over-all behaviour. The two transitions between successive processes cause some difficulties, particularly the one between melting and melt flow, where the melting models described here become increasingly unrealistic. In attempting to analyse machine performance, the mass flow rate along the screw must normally be prescribed. It is this parameter that provides the link between solids conveying, melting and melt flow, and between successive channel cross sections within the regions associated with these processes. Provided the extruder is operating steadily, the mass flow rate is independent of both time and position along the screw. In practice, the flow rate is often determined by a balance between the performance of the screw and the flow characteristics of the breaker plate, screen pack, die and any other restrictions at the delivery end of the extruder. These characteristics, which can often be computed by the methods described in sections 5.1 and 5.2, provide relationships between the flow rates and the delivery pressures that must be generated by the screw. Given a flow rate, an analysis of over-all machine performance can, in principle, predict the pressure profile along the screw, and hence the delivery pressure. With the aid of an iterative procedure, the flow rate can be adjusted until the pressures balance.

The above mode of operation is often described as melt controlled, although this term has also been used in the more restricted sense of a balance being achieved between the melt flows in the die and metering section alone. As the melting process is capable of generating large pressure rises, its contribution to the over-all pressure balance cannot be neglected. Indeed, a common situation in practice is for the melting process to generate a higher pressure than that required at the delivery end of the screw. The axial pressure profile then shows a peak at or near the beginning of the metering section, the downstream pressure gradient in the metering section is negative, and the drag flow rate capacity is exceeded.

The contribution of the solids conveying process is generally not important in melt-controlled operation: the practical example given in section 6.2.1 shows that the actual mass flow rate is usually well within the feeding capacity of the screw. The other main mode is that of feed-controlled operation, where the flow rate is determined by the capacity of either the hopper supplying the extruder or the feed section of the screw. An accurate analysis of solids conveying is clearly much more important in such a situation. In the case of melt-fed machines, feed-controlled operation appears to be much more common, and an example is discussed in section 6.7.4.

6.7.2 Design Criteria

With the aid of analyses of extruder performance, it is possible to study a given machine processing a particular polymer under known operating conditions. The next step towards improving the design or performance of this machine is to investigate the effects of changing, for example, the screw dimensions or the operating conditions. To do this effectively, appropriate design criteria must be prescribed. Although the relative importance of these vary according to the application, they are likely to include at least some of the following: (1) maximum output; (2) dimensional uniformity of the extrudate; (3) adequate mixing; (4) delivery temperature as uniform as possible and within prescribed limits; (5) minimum machine size and cost; (6) minimum power consumption; (7) maximum machine life. Such requirements, which are concerned with maximising quantity and quality while minimising capital and running costs, are of course very interdependent. Given the design constraints, it is possible, in principle, to determine an optimum design and set of operating conditions. An example concerning large hot-melt extruders is outlined in section 6.7.4.

6.7.3 Extruder Scale-up

An important use for analyses of extruder performance is in predicting the behaviour of an extruder larger than those previously employed for a particular application. For example, having established and possibly optimised a process on laboratory scale equipment, it is often required to design larger versions for production purposes. A rational way to attempt to do this is with the aid of dimensional analysis. Considering melt flow in the metering section of an extruder, it is clear from sections 4.5 and 6.1.1 that the important characteristic flow parameters are the dimensionless downstream pressure gradient, Griffith number and Graetz number (in preference to the Peclet number). In order to maintain similar flow conditions during scale-up, the values of these parameters should be unchanged. Note that if both the Griffith number and the set barrel temperatures are unchanged, then so is the Brinkman number, which characterises the thermal boundary conditions. For present purposes, it is assumed that dimensionless geometric parameters, such as helix angle and length-to-diameter ratios of the three screw sections, remain unchanged.

The purpose of the following analysis is to determine how the screw speed, N, and channel depth, represented by the metering-section depth, H, should be varied with increasing screw diameter, D. Now, the characteristic velocity, shear rate and shear stress vary as follows

$$V_z \sim ND, \qquad \bar{\gamma} \sim \frac{ND}{H}, \qquad \bar{\tau} \sim \left(\frac{ND}{H}\right)^n \qquad (6.131)$$

Assuming the delivery pressure to be unchanged, $P_z \sim D^{-1}$, and for a given polymer, equations 6.29, 6.31 and 6.32 imply that

$$\pi_P \sim D^{-1}\frac{H}{\bar{\tau}} \sim \frac{Gz}{G} \sim 1 \qquad (6.132)$$

$$G \sim \left(\frac{ND}{H}\right)^{n+1} H^2, \qquad Gz \sim NH^2 \tag{6.133}$$

Equation 6.132 shows that if delivery pressure, Gz and G are unchanged during scale-up, then so is the dimensionless pressure gradient. For the last two conditions to hold, it can be shown from equations 6.133 that

$$H \sim D^{(n+1)/(3n+1)}, \qquad N \sim D^{-2(n+1)/(3n+1)} \tag{6.134}$$

As the dimensionless flow rate also remains constant, equation 6.34 implies that

$$Q \sim WHV_z \sim D^2 HN \sim D^{(5n+1)/(3n+1)} \tag{6.135}$$

Therefore, for a typical value of $n = 0.4$

$$H \sim D^{0.64}, \qquad N \sim D^{-1.27}, \qquad Q \sim D^{1.36} \tag{6.136}$$

Although the above treatment is concerned only with melt flow, essentially identical results can be obtained from dimensional analyses of solids conveying and melting. The practical implications of equation 6.136 are that for similar flow conditions, in particular thermal conditions, to be maintained during scale-up, both the relative channel depth, H/D, and to a much greater extent the screw speed must be decreased, and the resulting output would be not much more than linearly proportional to machine size. The fact that, in practice, for economic reasons outputs are increased in proportion to at least the square of the diameter implies that similar flow conditions are not maintained, which helps to explain the difficulties often encountered with large machines. Both G and Gz usually increase with extruder size, and temperature variations become more severe, as illustrated in the following case study. While strict flow similarity is rarely, if ever, maintained during scale-up, quality criteria including adequate melting and mixing provide the design constraints.

6.7.4 Large Extruder Design Study

The following practical design study for a large hot melt-fed low-density polyethylene homogeniser serves a number of purposes, including to emphasise the thermal characteristics of large extruders, to provide a further illustration of the developing melt flow analysis described in section 6.1.4 and to demonstrate the use of mixing and power consumption calculations. Suppose that it is required to homogenise a low-density polyethylene of melt flow index 0.2 (see section 3.3.2), supplied at a rate of 5000 kg/h and a uniform temperature of 250 °C. The properties of this material are

$$\mu_0 = 16.0 \text{ kN s/m}^2 \text{ at } \gamma_0 = 1 \text{ s}^{-1} \text{ and } T_0 = 250 °C, n = 0.30$$

$$b = 0.01 °C^{-1}, \rho = 750 \text{ kg/m}^3, C_p = 3400 \text{ J/kg} °C, k = 0.30 \text{ W/m} °C$$

Certain assumptions are made concerning the extruder-screw geometry — including $\theta = 17.66°$, $e = 0.1D$, where D is the barrel diameter, and the lengths of the feed and compression sections are each $2D$. The compression ratio between the channel depths of the feed and metering sections is chosen to be 4, for the following reason. The output rate from a melt-fed extruder is normally determined by the feed section, where practical experience shows the dimensionless flow rate to be very close to 0.10. A value of π_Q in the metering section of about 0.4 is appropriate, being close to the maximum associated with drag flow there.

Initially one set of machine geometry is considered: $D = 300$ mm, over-all length $L = 25D$, and $H = 12$ mm in the metering section. For $\pi_Q = 0.4$ there, the screw speed may be determined as $N = 97$ rev/min. The barrel temperature is set at $T_b = 250\ °C$, with a view to maintaining the temperature of the supplied melt. Dimensionless flow parameters for the metering section are

$$Re = 2.34 \times 10^{-2}, Pe = 1.48 \times 10^5, G = 39.2, Gz = 85.5$$

implying strongly nonisothermal, convection dominated flow.

The performance predicted by the developing melt flow analysis is as follows

Bulk mean delivery temperature, $\overline{T}_{out}$ = 326 °C

Mechanical power consumption   = 690 kW

Barrel cooling rate   = 200 kW

Mean degree of mixing, $\overline{M}$ = 4400

Mean metering section shear rate, $\overline{\gamma}$ = 121 s^{-1}

As the solution is a numerical one, mixing is computed using equation 6.128, rather than the much simpler analytical expression displayed in equation 6.129. The power consumption is calculated with the aid of equation 6.122, ignoring leakage flow. Note that barrel cooling is required to maintain the required surface temperature, but that the bulk mean temperature still rises by 76 °C (and local temperatures within the flow by over 90 °C). In view of the large Graetz number, it is not surprising that the flow at the delivery end of the screw is far from being fully developed, which is fortunate because the temperatures associated with such flow would cause thermal degradation of the polymer.

Now, suppose that the effects of changing, say, barrel temperature, channel depths and screw diameter are to be studied. The first of these is shown in figure 6.18, in which the predicted bulk mean delivery temperature, mechanical power input and barrel cooling are plotted. The degree of distributive mixing is not significantly affected by barrel temperature. According to figure 6.18, the delivery temperature actually decreases as the barrel temperature is increased (eventually reaching a minimum). This serves to emphasise that in large extruders the heat generated by dissipation of mechanical work is mostly convected downstream, causing a continual rise in temperature, rather than conducted out through the barrel. The power input required is very sensitive to barrel temperature, and the main effect of barrel cooling is to draw power from the motor rather than heat from the polymer. The common practice of quoting specific power consumption (for example, 0.138 kW h/kg for the above case of $T_b = 250\ °C$) as a measure of mixing can be misleading, because here the true degree of mixing is not changing.

Figure 6.19 shows the effects of varying screw diameter for a fixed L/D ratio of 25. For each of the diameters, there is nearly a direct proportionality between mean temperature rise and degree of mixing, with metering-section channel depth

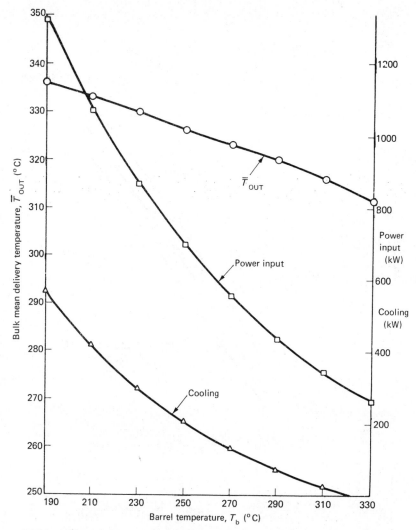

Figure 6.18   *Effects of varying barrel diameter for D = 300 mm, L/D = 25,*
H = 12 mm, π_Q = 0.4 (N = 97 rev/min)

as a variable. For a given value of the ratio H/D, the degree of mixing is virtually
constant. The proportionality between temperature rise and mixing is a feature of
large homogenisers: improved mixing can only be achieved at the expense of a
hotter melt. An important point to note from figure 6.19 is that machine
performance is improved by increasing the screw diameter, because a cooler melt
can be achieved for a given level of mixing. Considering the relatively massive
increases in extruder dimensions, however, the temperature reductions are modest.
Similar very modest improvements can be achieved by increasing the L/D ratio,
although for reasons of ease and cost of manufacture this is undesirable. Hence,

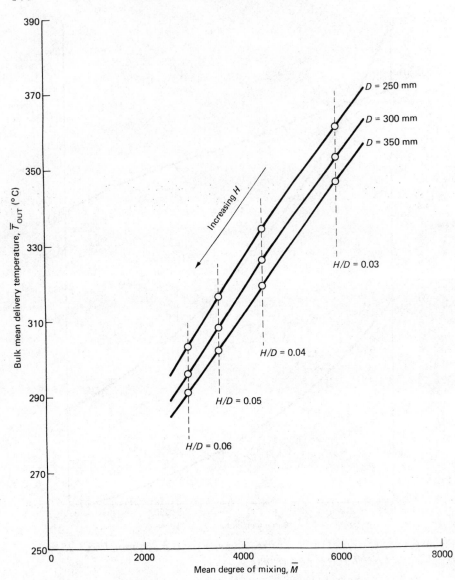

Figure 6.19   *Effects of varying screw diameter and relative channel depths for*
L/D = 25, π_Q = 0.4, T_b = 250 °C

D = 300 mm and L/D = 25 is a reasonable compromise for the present application.
The final choice of channel depth must depend on the mixing performance
required. If necessary this must be established by either laboratory trials or past
experience: in this case, an $\bar{M}$ in excess of 4000 is appropriate. Hence, the original
choice of H = 12 mm, which also corresponds to a screw speed and mean shear rate
that are not excessive in terms of wear and shear degradation.

7

Injection Moulding

The injection-moulding process was briefly described in section 2.2. Unlike extrusion, which is, or should be, a continuous steady process, injection moulding is discontinuous, although it normally follows a regularly repeated cycle of operations. During injection, the flow in a mould cavity is clearly time dependent, in that one flow boundary moves as the material advances to fill the mould. Also, in analyses of any parts of the process, it may not be permissible to ignore time-derivative terms in the equations governing flow and heat transfer. Another important feature is that injection moulding involves very severe flow conditions in terms of both pressures and rates of deformation. Shear rates achieved during injection are often higher than those for which viscosity data can be conveniently and reliably obtained. Pressures are often large enough to cause significant changes in material properties, particularly melt viscosity.

For the purposes of analysis, the injection-moulding process can be broken down into at least three components. These are the melting or *plastication* achieved by a screw or other device, flow in the heated injection nozzle and flow and heat transfer in the cold mould passages and cavities.

7.1 Reciprocating-screw Plastication

In a screw injection-moulding machine, melting of the polymer prior to injection is achieved in much the same way as in an extruder. Thus, the melting mechanism in a single-screw machine is similar to that described in section 6.3.1. Indeed, if screw rotation and axial motion, which is often referred to as *screwback* in the present context of reciprocating screw injection-moulding machines, occupy a large proportion of the total moulding-cycle time, the screw melting performance is essentially that of the same machine extruding continuously. There are, however, some significant differences when, as often happens, screwback occurs for relatively short periods of time. A substantial amount of melting of material from the solid bed then occurs by thermal conduction while the screw is stationary. During screwback, there is considerable readjustment of the position of the solid bed in the screw channel. Although the screw moves axially while it is rotating, the axial velocity component is not large enough to influence melting behaviour significantly.

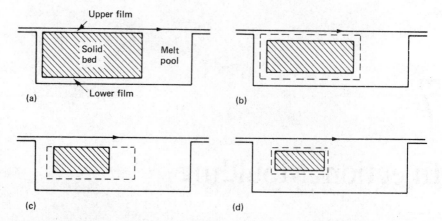

Figure 7.1   *The melting mechanism in reciprocating single-screw plastication*

This is not so true, however, of the injection stroke, when the axial velocity of the screw is its only component.

Figure 7.1 shows diagrammatically the changes in solid bed and melt that may be observed experimentally at a particular screw-channel cross section during a typical plastication and injection cycle. Figure 7.1a shows the situation at the end of screwback, which is similar to that obtained in continuous extrusion with an uncooled screw (figure 6.14b). During the soak period prior to injection, the screw is stationary and thermal conduction melts material from all sides of the solid bed, as shown by figure 7.1b. Then, while the screw is moving forward to inject melt into the mould, some further melting in the screw channels occurs both by thermal conduction and as a result of the motion of the screw. Although the thickness of the upper melt film appears to remain largely unchanged, and considerably larger than during screwback, the bed reduces significantly in both width and depth as shown in figure 7.1c. It is the axial motion of the screw during injection that controls the thickness of the upper melt film and contributes to the reduction of the solid bed width, by the usual mechanism of sweeping melt into the melt pool. During the hold period, when injection pressure on the mould is maintained, there is some further melting by conduction, as illustrated in figure 7.1d. Finally, during screwback, downstream motion of the tapering solid bed relative to the screw serves to increase the proportion of solid present at any channel cross section, until the situation shown in figure 7.1a is regained.

In principle, the melting-model analysis developed for single-screw extruders can be adapted to allow for the various stages of reciprocating-screw plastication. In practice, however, there is less incentive to do this because screw design for injection-moulding machines is less critical than for extruders. Such experiments and analysis as have been carried out show that the polymer is almost invariably not completely molten when it leaves the relatively short screw of an injection-moulding machine, and that solid-bed break-up usually occurs. The surging associated with the latter does not seriously affect the over-all moulding process. Incomplete melting is remedied either by thermal conduction prior to injection, or by the intense shear imparted when the material is forced through the nozzle and mould passages.

7.2 Melt Flow in Injection Nozzles

The flow behaviour of a polymer melt in the nozzle of an injection-moulding machine can have a considerable effect on the over-all process. A substantial proportion of the total injection pressure generated by the axial load on the screw may be absorbed in the nozzle. Also, as a result of the shear induced, melt temperatures increase sufficiently to affect the flow properties of the material entering the mould, and in extreme cases may lead to thermal degradation.

Injection nozzles are generally circular in cross section. Flow analysis of the type discussed in section 5.1.1 for extrusion dies of similar shape is appropriate, provided due allowance is made for unsteady flow and heat transfer, and also for the influence of pressure on melt properties. Before considering particular solutions, it is appropriate to apply dimensional analysis to establish the type of flow regime involved.

7.2.1 Dimensionless Parameters for Nozzle Melt Flow

Consider, for example, an injection nozzle of the form shown in figure 7.2, which is a standard design used for testing moulding machine performance. The corresponding standard mass flow rate is 50 g/s. Most of the pressure drop and heat transfer occur in the final narrow parallel region and so, for the purposes of dimensional analysis, the nozzle can be regarded as a circular passage of diameter $D = 2.38$ mm and length $L = 25.4$ mm. Suppose that a low-density polyethylene of melt-flow index 0.4, initially at a temperature of 140 °C, is injected through this nozzle. The melt properties at ambient pressure and temperatures of the order of 140°C are

$$\mu_0 = 30.8 \text{ kN s/m}^2 \text{ at } \gamma_0 = 1 \text{ s}^{-1} \text{ and } T_0 = 140 \text{ °C}, n = 0.337$$

$$b = 0.0137 \text{ °C}^{-1}, \rho = 775 \text{ kg/m}^3, C_p = 2540 \text{ J/kg °C}, k = 0.24 \text{ W/m °C}$$

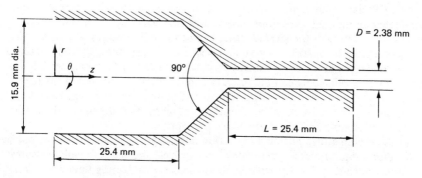

Figure 7.2   *Cross section of a standard injection nozzle*

For a mass flow rate of 50 g/s, the volumetric flow rate is

$$Q = \frac{50 \times 10^{-3}}{775} = 6.45 \times 10^{-5} \text{ m}^3/\text{s}$$

and the characteristic mean velocity is

$$\bar{U} = \frac{4Q}{\pi D^2} = 14.5 \text{ m/s}$$

Taking the nozzle diameter, D, as the characteristic distance between the flow boundaries, the mean shear rate, viscosity and shear stress are

$$\bar{\gamma} = \frac{\bar{U}}{D} = 6090 \text{ s}^{-1}$$

$$\bar{\mu} = \mu_0 \left| \frac{\bar{\gamma}}{\gamma_0} \right|^{n-1} = 30.8 \times 10^3 \times 6090^{-0.663} = 95.3 \text{ N s/m}^2$$

$$\bar{\tau} = \bar{\mu}\bar{\gamma} = 581 \text{ kN/m}^2$$

The characteristic Reynolds, Peclet, Griffith and Graetz numbers for the flow, defined according to equations 4.60, 4.62, 4.63 and 4.65, are

$$Re = \frac{\rho \bar{U} D}{\bar{\mu}} = \frac{775 \times 14.5 \times 2.38 \times 10^{-3}}{95.3} = 0.281$$

$$Pe = \frac{\rho C_p \bar{U} D}{k} = \frac{775 \times 2540 \times 14.5 \times 2.38 \times 10^{-3}}{0.24} = 2.83 \times 10^5$$

$$G = \frac{b\bar{\tau}\bar{\gamma} D^2}{k} = \frac{0.0137 \times 581 \times 10^3 \times 6090 \times (2.38 \times 10^{-3})^2}{0.24} = 1140$$

$$Gz = Pe \frac{D}{L} = 2.83 \times 10^5 \times \frac{2.38}{25.4} = 2.65 \times 10^4$$

The comparatively large size of the Reynolds number implies that inertia effects, although small, are not as negligible as in most melt flows. With such enormous values of Peclet number, and particularly Graetz number, heat transfer is entirely dominated by thermal convection: the flow is effectively adiabatic, with negligibly small amounts of heat transferred either to or from the surrounding nozzle. Also, the flow leaving the nozzle is a very long way from being thermally fully developed. Therefore, although the exceptionally large value of the Griffith number implies very strong coupling between the velocity and temperature profiles in the highly nonisothermal flow once a significant degree of thermal development has taken place, the almost total lack of this development means that the use of the isothermal approximation may provide a reasonable first approximation.

Two special features of flow in injection nozzles are the transient behaviour due to the discontinuous nature of the moulding process and the very high pressures involved, which significantly affect melt viscosities. Considering time dependence first, in the above practical example, the mean residence time in the nozzle is

$L/\bar{U} = 1.75$ ms. Time-derivative terms, which represent inertia and convection effects in the momentum and energy equations, respectively, are only likely to be significant in the energy equation. Because thermal convection dominates the heat transfer, it is reasonable to expect that transient effects will have disappeared after a period equal to a comparatively small number of residence times. In the numerical example, this period is unlikely to exceed, say, 10 ms, a conclusion confirmed by much more detailed calculations. The important point, however, is that this time is negligibly small compared with the injection time, which is likely to be several seconds, and the flow may be regarded as steady.

As for pressure effects on viscosity, equation 3.20 defines a convenient empirical form of dependence. A dimensionless pressure-dependence parameter can therefore be defined as

$$\phi = \alpha \Delta P \qquad (7.1)$$

where α is the pressure dependence of viscosity, and ΔP is the pressure difference across the nozzle. Note that it is pressure difference rather than absolute pressure that is important, because it determines the degree of variation of viscosity within the flow due to pressure effects. If pressures are high throughout the flow, it is merely necessary to increase the general level of viscosities accordingly. The effects of pressure dependence can be expected to be significant if the parameter ϕ is of the order of 0.1 or more, when, owing to the exponential form of equation 3.20, the variation in viscosity is over 10 per cent. The polyethylene considered in the above practical example has an α of about 7.3 m^2/GN. The value of ΔP necessary to determine ϕ can be estimated by a more detailed analysis of the nozzle flow.

7.2.2 Analysis of Nozzle Melt Flow

The simplest form of nozzle flow analysis is that described in section 5.1.1 for isothermal melt flow in an extrusion die of circular cross section. As concluded from the discussion of dimensionless nozzle flow parameters in the above practical example, the use of the isothermal approximation may provide a reasonable first approximation, at least for estimating pressure gradients. From equation 5.11, the dimensionless pressure gradient for the particular polyethylene concerned is

$$\pi_P = \frac{P_z D}{\bar{\tau}} = -2^{2.337} \left(\frac{2.011}{0.337} \right)^{0.337} = -9.22$$

where P_z is the pressure gradient along the nozzle, which is

$$P_z = \frac{\pi_P \bar{\tau}}{D} = \frac{-9.22 \times 581}{2.38} = -2250 \text{ MN/m}^3$$

The over-all pressure difference may therefore be estimated as

$$\Delta P = -P_z L = 2250 \times 25.4 \times 10^{-3} = 57.2 \text{ MN/m}^2$$

which absorbs a substantial proportion of the total injection pressure (typically 80–200 MN/m^2). With this estimated pressure difference, the pressure dependence parameter is

$$\phi = \alpha \Delta P = 7.3 \times 10^{-3} \times 57.2 = 0.417$$

implying a more than 50 per cent variation in viscosity. Clearly, the pressure dependence of viscosity is a significant factor in this case.

The pressure dependence can be readily included in an isothermal analysis of nozzle flow in which pressure varies only with axial position. The axial pressure gradient may be obtained from equation 5.11 as

$$\frac{dp}{dz} = P_z = \frac{\pi_P \bar{\tau}}{D} \tag{7.2}$$

where π_P is a known function of the power-law index and $\bar{\tau}$ is the local value of the mean shear stress. If $\bar{\tau}_0$ is the value $\bar{\tau}$ would take at ambient pressure, then by virtue of the exponential dependence of viscosity on pressure defined in equation 3.20

$$\bar{\tau} = \bar{\tau}_0 \exp(\alpha p) \tag{7.3}$$

where p is the local pressure. Therefore

$$\frac{dp}{dz} = \frac{\pi_P \bar{\tau}_0}{D} \exp(\alpha p) \tag{7.4}$$

where π_P, $\bar{\tau}_0$ and D are independent of axial position. This differential equation can be integrated to give

$$\exp(-\alpha p) = -\alpha \frac{\pi_P \bar{\tau}_0}{D} z + A \tag{7.5}$$

where A is an integration constant. Applying the boundary conditions $p = p_1$ at $z = 0$, the nozzle inlet, and $p = p_2$ at $z = L$, the nozzle exit

$$\exp(-\alpha p_2) - \exp(-\alpha p_1) = -\alpha \frac{\pi_P \bar{\tau}_0}{D} L = \phi \tag{7.6}$$

where ϕ is determined as above for a pressure difference calculated without allowing for pressure-dependence effects. Returning to the numerical example, in which $\phi = 0.417$, if the inlet pressure is $p_1 = 100$ MN/m^2, equation 7.6 yields for p_2 and the pressure difference

$$\exp(-7.3 \times 10^{-3} \times p_2) = 0.417 + \exp(-7.3 \times 10^{-3} \times 100)$$

$$p_2 = 14.5 \text{ MN/m}^2, \qquad p_1 - p_2 = 85.5 \text{ MN/m}^2$$

On the other hand, if p_1 is increased to 200 MN/m^2, the pressure difference rises to 141 MN/m^2, a somewhat smaller proportion of the available injection pressure. Clearly, the pressure dependence of viscosity can, in principle, have a very significant effect on flow at the pressures involved in injection moulding.

In practice, however, this effect is often to a large extent offset by temperature rises due to the intense viscous dissipation, which reduce viscosities and hence pressure gradients. Following the very simple analysis introduced in section 3.3.3, an order of magnitude for the mean temperature rise in the nozzle can be estimated from equation 3.18 as

$$\Delta T = \frac{\Delta P}{\rho C_p} \qquad (7.7)$$

where ΔP is the pressure difference calculated without allowing for pressure-dependence effects. In the numerical example

$$\Delta T = \frac{57.2 \times 10^6}{775 \times 2540} = 29.1 \, ^\circ C$$

which implies about a 50 per cent variation in viscosity. This similarity between pressure- and temperature-dependence effects is due to the fact that, for the polyethylene concerned, the ratio

$$\frac{b}{\rho C_p} = \frac{0.0137}{775 \times 2540} = 6.96 \times 10^{-9} \, m^2/N$$

is very close to the value of α, and therefore

$$b\Delta T \approx \alpha \Delta p \qquad (7.8)$$

This balance is, of course, only relevant in pressure flows. Note that the estimated mean temperature rise is substantial, despite the fact that the degree of thermal development achieved in the nozzle is extremely small. Clearly, the maximum temperatures in fully developed flow would be enormous.

A thorough analysis of flow and heat transfer in the nozzle of an injection-moulding machine should allow for development with respect to both time and position along the nozzle, and also for the influences of temperature and pressure on viscosity. This can and has been done by a method of analysis broadly similar to that described in section 6.1.4 for developing melt flow in an extruder screw channel. One of the difficulties of applying such an analysis is to select realistic thermal boundary conditions, without simultaneously analysing heat transfer in the metal of the nozzle itself. In view of the intense and transient nature of the flow, it is unreasonable to prescribe fixed boundary temperatures, which must rise during injection. Similarly, the assumption of zero radial temperature gradients at the boundaries in recognition of the adiabatic nature of the flow is not entirely justified, but appears to be the more realistic of the two.

Application of a developing flow analysis to the practical nozzle flow example considered previously confirms a number of the conclusions already drawn — namely, that the flow becomes steady after a few milliseconds, and that the effects of temperature and pressure on viscosity tend to cancel out in terms of pressure predictions. The simple isothermal analysis of nozzle flow gives reasonably satisfactory estimates of pressure differences. The detailed analysis of temperatures does show, however, that temperature rises are far from uniform across the flow and are greatest in the region of high shear rate near the nozzle wall. In other words, a form of plug flow is created, with a relatively cool core of material lubricated by a hot low-viscosity layer at its boundaries.

7.3 Flow and Heat Transfer in Moulds

The mould filling process is an extremely complex one in terms of melt flow, heat transfer and solidification, which lead to molecular orientation, residual stresses and shrinkage in the finished moulding. Because of this complexity, only comparatively modest advances have been made in the analysis of the process. Attention is confined here to a qualitative discussion of the mechanism of mould filling, together with some simple analyses of the flow and heat transfer involved. Flow in cold sprues, runners and gates is similar to that in mould cavities, in that solidification at the flow boundaries takes place.

7.3.1 Mechanism of Flow in a Mould Cavity

As indicated in section 2.2.3, provided the rate of injection is not so high as to produce jetting of melt through the gate and across the mould cavity, the flow spreads outward away from the gate, with a core of hot melt flowing between solidified skins formed on the cavity surfaces. Figure 7.3 shows a diagrammatic cross-section of a mould cavity through an advancing flow front. At this front, hot melt from the centre flows outwards to meet the cold cavity walls, where it immediately solidifies. Subsequent growth of skin thickness is relatively slow, and the skins remain thin during the remainder of the injection part of the moulding cycle.

For the purposes of analysis, flow and heat transfer in a mould cavity can be considered in two stages. These are melt flow between solidified skins during injection, and thermal conduction during the hold and cooling parts of the moulding cycle. During the hold period, it is reasonable to neglect the small amount of melt flow that occurs in order to keep the cavity full as the polymer cools and contracts.

7.3.2 Dimensionless Parameters for Mould Cavity Flow

The dimensionless parameters characterising melt flow in a mould cavity vary widely according to the geometry of the cavity, and very often according to

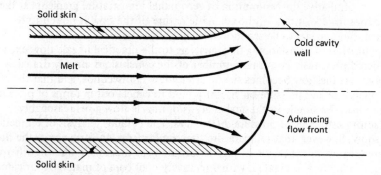

Figure 7.3   *Mechanism of filling a mould cavity*

position within the cavity as well. In many mould arrangements, injected melt spreads outwards from a small gate. Therefore, while the flow conditions are intense near the gate, they are much less so at the flow front when the cavity is nearly full. As an illustration, consider the material and injection rate defined for the nozzle flow example introduced in section 7.2.1. Suppose the melt is advancing along a flow front of length 100 mm in a relatively deep mould cavity of depth $H = 3$ mm. For the injection rate of $Q = 6.45 \times 10^{-5}$ m^3/s, the mean velocity at the front, which is inversely proportional to the length of this front, is

$$\bar{U} = \frac{6.45 \times 10^{-5}}{3 \times 10^{-3} \times 100 \times 10^{-3}} = 0.215 \text{ m/s}$$

which is much lower than the mean nozzle velocity. Taking the depth H as the characteristic distance between the flow boundaries, the mean shear rate can be found as $\bar{\gamma} = \bar{U}/H = 71.7$ s^{-1}. The main dimensionless parameters at the flow front may be calculated as

$$Re = 2.76 \times 10^{-4}, Pe = 5290, G = 4.79$$

Similarly, for a flow front of the same length in a relatively shallow cavity of depth $H = 1$ mm, the Peclet number is unchanged while the Griffith number increases to 10.0. Thus, Pe is generally of the order of 10^3 or more, while G is often in the range 1–10. Thermal convection is still the dominant mode of heat transfer, and velocity and temperature profiles are fairly strongly coupled, although this coupling is much weaker than in nozzle flow.

7.3.3 Analysis of Mould Filling

As a first approximation, the growth of the solid skins during mould filling may be assumed to be negligible, and the effective flow channel dimensions are those of the cavity itself. Also, the flow is assumed to be both locally steady, in that time derivatives in the governing equations are ignored, and isothermal. Clearly, the flow is time dependent in the sense that the flow front moves, but some justification for ignoring time derivatives was provided in section 7.2.1. Although the isothermal assumption is much less satisfactory, it does allow relatively simple algebraic formulae to be derived for use in mould design calculations. Although calculated flow rates or pressures may be reasonably satisfactory, more sophisticated nonisothermal flow analyses must be undertaken if detailed predictions of velocity and temperature profiles are required.

Some simple mould filling situations are illustrated in figure 7.4. They involve cavities of either circular or flat slit cross-section, containing either one-dimensional or essentially axisymmetric radial melt flows. Isothermal analyses can be applied to each case to establish relationships between flow rate and pressure requirements.

Figure 7.4a shows a cavity of circular cross-section with diameter D. From equation 5.11, the pressure gradient at any position along the cavity is given by

$$\frac{dp}{dz} = P_z = \frac{\pi_p \bar{\tau}_0}{D} \left(\frac{Q}{Q_0}\right)^n \left(\frac{D_0}{D}\right)^{3n} \tag{7.9}$$

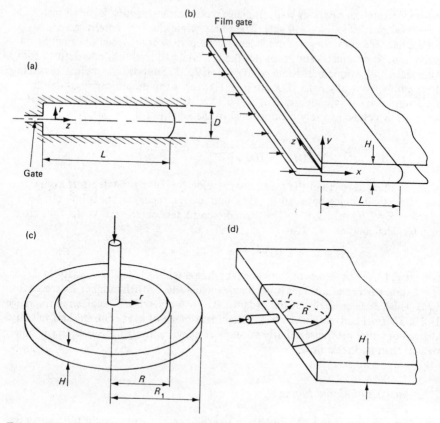

Figure 7.4   *Some simple mould cavities and methods of filling*

where π_P is negative and a known function of power-law index, and $\bar{\tau}_0$ is the mean shear stress defined according to equations 5.10 for a volumetric flow rate Q_0 in a cavity of diameter D_0 at the appropriate melt temperature. The reason for introducing the reference values Q_0 and D_0 is to make it possible to study the effects of varying flow rate and cavity diameter. Note the power-law form of the dependence of pressure gradient on Q, which agrees with equation 5.9, where Q is the volumetric flow rate into the mould at a particular time during the filling process. Since the pressure is zero at the flow front, the pressure required just after the gate is

$$P = -P_z L = \frac{(-\pi_P)\bar{\tau}_0}{D_0} \left(\frac{Q}{Q_0}\right)^n \left(\frac{D_0}{D}\right)^{3n+1} L \tag{7.10}$$

where L is the length of the cavity filled. Now, if Q is constant and equal to Q_0, this length is the following linear function of time, t

$$L = \frac{4Q_0 t}{\pi D^2} \tag{7.11}$$

and

$$P = \frac{4(-\pi_P)\bar{\tau}_0 Q_0 t}{\pi D_0{}^3} \left(\frac{D_0}{D}\right)^{3n+3} \qquad (7.12)$$

Note that the pressure required is also a linear function of time, and is extremely sensitive to the diameter of the cavity.

Although constant injection rate is the more common condition used in practice, an alternative is to maintain constant injection pressure. This condition can be simulated by assuming constant gate pressure, P, and rearranging equation 7.10

$$\frac{Q}{Q_0} = \left[\frac{PD}{(-\pi_P)\bar{\tau}_0 L}\right]^{1/n} \left(\frac{D}{D_0}\right)^3 = \left(\frac{L_0}{L}\right)^{1/n} \left(\frac{D}{D_0}\right)^{1/n+3} \qquad (7.13)$$

where L_0 is the length filled with the same gate pressure when the flow rate is Q_0 in a cavity of diameter D_0. The rate of filling of the actual cavity is given by

$$\frac{dL}{dt} = \frac{4Q}{\pi D^2} = \frac{4Q_0}{\pi D_0{}^2} \left(\frac{L_0}{L}\right)^{1/n} \left(\frac{D}{D_0}\right)^{1/n+1} \qquad (7.14)$$

Integration of this differential equation with the initial condition $L = 0$ at $t = 0$ gives

$$\left(\frac{L}{L_0}\right)^{1/n+1} = \frac{4Q_0 t}{\pi D_0{}^2 L_0} \left(\frac{D}{D_0}\right)^{1/n+1} \frac{(n+1)}{n} \qquad (7.15)$$

In other words, the length filled is proportional to the $n/(n+1)$th power of the time elapsed. For a newtonian melt, L is therefore proportional to the square root of t, while for a more practical value of, say, $n = 0.5$, it is proportional to the cube root. Note that the length is also directly proportional to cavity diameter, irrespective of the power-law index of the melt. From equations 7.13 and 7.15 the variation of injection rate with time can be deduced, in the form

$$Q \propto t^{-1/(n+1)} \qquad (7.16)$$

Comparable forms of expressions can be obtained for the flat slit cavity fed from a film gate shown in figure 7.4b, important differences being that the form of π_P defined by equation 5.25 is used in place of that given in equation 5.11, and the cavity depth H replaces D as the distance between the flow boundaries. The radial flows shown in figures 7.4c and 7.4d are very similar to each other, the latter being essentially half the former. Consider the complete disc shown in figure 7.4c, in which the flow can be treated as flow in the radial direction in a narrow slit. A good approximation for the radial pressure gradient may therefore be obtained from equation 5.25 as

$$\frac{dp}{dr} = P_r = \frac{\pi_P \bar{\tau}}{H} \qquad (7.17)$$

where H is the depth of the mould cavity, π_P is a known function of the power-law index, and $\bar{\tau}$ is the local mean shear stress. Now, this stress is defined at the appropriate melt temperature and the local mean shear rate based on the mean radial velocity, which is inversely proportional to the radial distance from the injection point

$$\bar{U} = \frac{Q}{2\pi r H}, \qquad \bar{\gamma} = \frac{\bar{U}}{H}, \qquad \bar{\tau} = \mu_0 \left| \frac{\bar{\gamma}}{\gamma_0} \right|^{n-1} \bar{\gamma} \qquad (7.18)$$

where Q is the total volumetric flow rate into the mould at a given time. Hence, the radial pressure gradient may be expressed as

$$P_r = \frac{\pi_P \bar{\tau}_0}{H} \left(\frac{Q}{Q_0} \right)^n \left(\frac{r_0}{r} \right)^n \left(\frac{H_0}{H} \right)^{2n} \qquad (7.19)$$

where $\bar{\tau}_0$ is the mean shear stress at some reference radius, r_0, when the flow rate is Q_0 in a cavity of depth H_0. Now, the pressure required just after the gate is

$$P = -\int_0^R P_r \, dr = \frac{(-\pi_P)\bar{\tau}_0}{H_0} \left(\frac{Q}{Q_0} \right)^n \left(\frac{H_0}{H} \right)^{2n+1} \left(\frac{R}{r_0} \right)^{1-n} \frac{r_0}{(1-n)} \qquad (7.20$$

provided $n \neq 1$, where R is the radial distance of the flow front from the gate. The rate of change of this radius is given by

$$\frac{dR}{dt} = (\bar{U})_{r=R} = \frac{Q}{2\pi R H} \qquad (7.21)$$

Now, if the injection rate, Q, is constant and equal to Q_0, equation 7.21 may be solved for R

$$R = \left(\frac{Q_0 t}{\pi H} \right)^{1/2} \qquad (7.22)$$

and from equation 7.20

$$P = \frac{(-\pi_P)\bar{\tau}_0}{H_0} \left(\frac{H_0}{H} \right)^{(3n+3)/2} \left[\frac{1}{r_0} \left(\frac{Q_0 t}{\pi H_0} \right)^{1/2} \right]^{1-n} \frac{r_0}{(1-n)} \qquad (7.23)$$

Note that the pressure required varies as the $(1-n)/2$ power of time. If required, the radial pressure distribution and hence the total mould clamping force can also be obtained.

If the injection process is such that the gate pressure, P, is constant, then from equation 7.20

$$\frac{Q}{Q_0} = \left(\frac{R_0}{R} \right)^{(1-n)/n} \left(\frac{H}{H_0} \right)^{1/n+2} \qquad (7.24)$$

where R_0 is the flow front radius with the same gate pressure when the flow rate is Q_0 in a cavity of depth H_0. From equation 7.21, the rate of filling is given by

$$\frac{dR}{dt} = \frac{Q_0}{2\pi R_0 H_0} \left(\frac{R_0}{R} \right)^{1/n} \left(\frac{H}{H_0} \right)^{1/n+1} \qquad (7.25)$$

Integration with the initial condition $R = 0$ at $t = 0$ gives

$$\left(\frac{R}{R_0} \right)^{1/n+1} = \frac{Q_0 t}{2\pi R_0^2 H_0} \left(\frac{H}{H_0} \right)^{1/n+1} \frac{(n+1)}{n} \qquad (7.26)$$

which is similar in form to equation 7.15.

As an illustration of the use of the above disc injection formulae, consider the material and injection rate defined for the nozzle flow example introduced in section 7.2.1. Suppose the cavity is of depth $H = H_0 = 3$ mm, and outer radius $R_1 = 50$ mm. With an injection rate $Q = Q_0 = 6.45 \times 10^{-5}$ m^3/s, the filling time can be derived from equation 7.22 as

$$t = \frac{\pi R_1{}^2 H}{Q_0} = \frac{\pi \times (50 \times 10^{-3})^2 \times 3 \times 10^{-3}}{6.45 \times 10^{-5}} = 0.365 \text{ s}$$

Taking the reference radius, r_0, as 10 mm, the mean velocity, shear rate and shear stress for the melt temperature at which viscosity data were specified are given by equations 7.18 as

$$\bar{U} = \frac{Q}{2\pi r_0 H} = \frac{6.45 \times 10^{-5}}{2\pi \times 10 \times 10^{-3} \times 3 \times 10^{-3}} = 0.342 \text{ m/s}$$

$$\bar{\gamma} = \frac{\bar{U}}{H} = 114 \text{ s}^{-1}, \qquad \bar{\tau}_0 = \mu_0 \left| \frac{\bar{\gamma}}{\gamma_0} \right|^{n-1} \qquad \bar{\gamma} = 152 \text{ kN/m}^2$$

From equation 5.25, the dimensionless pressure gradient is given by

$$\pi_P = -2^{1.674} \left(\frac{1.674}{0.674} \right)^{0.337} = -4.34$$

and from equation 7.20, the maximum pressure required to just fill the mould is

$$P = \frac{4.34 \times 152 \times 10^{-3}}{3 \times 10^{-3}} \left(\frac{50}{10} \right)^{0.663} \frac{10 \times 10^{-3}}{0.663} = 9.64 \text{ MN/m}^2$$

For comparison, if the cavity depth is reduced to $H = 1$ mm and the outer radius increased to $R_1 = 100$ mm, the filling time increases slightly to 0.487 s but, in view of the relationship between gate pressure and cavity dimensions expressed in equation 7.20, the maximum pressure required increases to

$$P = 9.64 \times \left(\frac{3}{1} \right)^{1.674} \left(\frac{100}{50} \right)^{0.663} = 96.0 \text{ MN/m}^2$$

Suppose that the maximum gate pressure is restricted to 20 MN/m^2, and is maintained at this level during injection. According to equation 7.20, this would have been reached in the original 3 mm deep cavity with $Q = Q_0 = 6.45 \times 10^{-5}$ m^3/s at a radius, R_0, given by

$$\frac{20}{9.64} = \left(\frac{R_0}{50} \right)^{0.663}, \qquad R_0 = 150 \text{ mm}$$

Therefore, using equation 7.26, the injection time under the constant pressure conditions in the 1-mm-deep cavity is given by

$$\left(\frac{100}{150} \right)^{3.967} = \frac{6.45 \times 10^{-5} t}{2\pi(50 \times 10^{-3})^2 \times 3 \times 10^{-3}} \left(\frac{1}{3} \right)^{3.967} \frac{1.337}{0.337}$$

$$t = 2.88 \text{ s}$$

In the mould cavity filling flows illustrated in figure 7.4, the cavity geometry is sufficiently simple for the flow to be analysed with the aid of the algebraic formulae derived for simple extrusion die flows in section 5.1. In general, mould cavities are much more complex in shape, and often vary in thickness. It is still possible, however, to undertake isothermal flow analyses to establish flow rate and pressure relationships for an existing or proposed mould design. An appropriate method is the one described in section 5.2 for two-dimensional flows in the narrow channels of extrusion dies and crossheads.

7.3.4 Analysis of Mould Cooling

Once a mould cavity is full, cooling time must be allowed for the contents to solidify sufficiently to avoid warping when the finished moulding is ejected. The formation of solid skins at the surfaces of a typical flat moulding is illustrated in the cross-sectional view shown in figure 7.5. Cooling by thermal conduction through the thickness of the polymer can be analysed with the aid of a simplified form of energy equation 4.16

$$\rho C_p \frac{\partial T}{\partial t} = k \frac{\partial^2 T}{\partial y^2} \tag{7.27}$$

This equation should be applied to both the melt and solid layers, using the appropriate values of the material properties in each case. The main difficulty in trying to solve this equation, either analytically or numerically, is that the initial temperature profile in the melt is not normally known.

A useful lower bound on the cooling time can; however, be obtained by assuming that all the melt is close to the melting temperature, T_m, and confining the conduction analysis to the solid skins. Provided the rate of growth of skin thickness, h, is relatively slow, equation 7.27 is approximately satisfied by the following temperature profile in, say, the lower skin in figure 7.5

$$T = T_b + y \frac{(T_m - T_b)}{h} \tag{7.28}$$

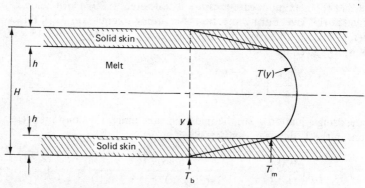

Figure 7.5   *Formation of solid polymer skins in a mould cavity*

where T_b is the temperature of the cavity boundary, which may not be known very accurately. Adapting equation 6.118, the rate of solidification per unit surface area of skin, ω, is given by

$$\omega\lambda = k_s \left(\frac{\partial T}{\partial y}\right)_s = \frac{k_s}{h}(T_m - T_b) \tag{7.29}$$

where λ is the latent heat of fusion and k_s is the thermal conductivity of the solid skin. But solidification increases the skin thickness

$$\omega = \rho_s \frac{dh}{dt} \tag{7.30}$$

where ρ_s is the density. Hence

$$\frac{dh}{dt} = \frac{k_s}{\rho_s\lambda}(T_m - T_b) \tag{7.31}$$

which may be integrated with the initial condition $h = 0$ at $t = 0$ to give

$$h^2 = \frac{2k_s t}{\rho_s\lambda}(T_m - T_b) \tag{7.32}$$

and skin thickness is proportional to the square root of the cooling time.

To illustrate the use of this result, consider a polymer with the following properties

$$\rho_s = 910 \text{ kg/m}^3, \, \lambda = 100 \text{ kJ/kg}, \, T_m = 110 \text{ °C}, \, k_s = 0.24 \text{ W/m °C}$$

Suppose the mould cavity is of depth $H = 3$ mm, and its surfaces are estimated to be at $T_b = 40$ °C. As the moulding will be completely solid when $h = H/2 = 1.5$ mm, the cooling time required can be estimated as

$$(1.5 \times 10^{-3})^2 = \frac{2 \times 0.24 \times 70}{910 \times 100 \times 10^3} t$$

$$t = 6.09 \text{ s}$$

This figure is likely to underestimate significantly the actual cooling time required. A more thorough analysis of mould cooling would need to take into account not only the initial melt temperature profile, but also the transient heat transfer through the metal of the mould to the cooling passages, and possibly further cooling of the polymer to well below its melting point.

Appendix A

Finite Element Analysis of Narrow Channel Flow

Although finite element methods were originally developed for digital computer use in the stress analysis of solid structures and components, they have also been applied to fluid mechanics and heat transfer problems, including the slow non-newtonian flows encountered in polymer processing operations. The method described here is designed to solve partial differential equations governing narrow channel flow, such as equation 5.55, together with equations 5.48 and 5.56, which specify the dependence of viscosity on the local state of the deformation rate. The use of this method is illustrated in section 5.2.2 in connection with flow in a cable covering crosshead deflector.

Figure 5.8 shows the solution domain in the x, z plane divided into triangular finite elements. Although the straight-sided element cannot follow the curved boundaries exactly, with a reasonable number of elements the maximum deviations are acceptably small. It should be noted that the number of elements across the width of the flow is constant, which means that the elements near the narrow inlet boundary are much smaller than those near the deflector outlet. It is the ability to fit complex geometric boundary shapes and to allow varying densities of elements within the solution domain that makes finite element methods attractive. While figure 5.8 shows only 72 elements covering the region of interest, rather more than this would be needed to obtain adequate accuracy. The form of distribution is, however, appropriate.

A variational approach can be used to solve the governing partial differential equation by seeking a stationary value for the functional χ, which is defined by an appropriate integration of the unknowns over the solution domain. It can be shown that the required stationary condition is obtained when

$$\frac{\partial \chi}{\partial \eta} = \int \int \frac{\bar{\mu}}{2H^3} \frac{\partial}{\partial \eta} \left[\left(\frac{\partial \psi}{\partial x} \right)^2 + \left(\frac{\partial \psi}{\partial z} \right)^2 \right] \mathrm{d}x \mathrm{d}z = 0 \qquad (A.1)$$

holds for all of the unknowns, η, required to be found. In the present method, the values of the stream function, ψ, at the corners or nodes of all the triangular elements are chosen as the unknowns. The only restriction on the validity of

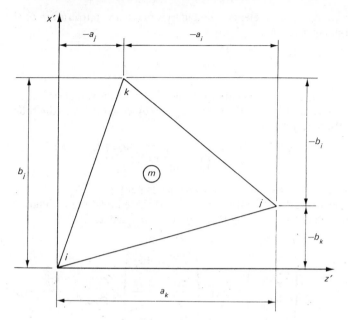

Figure A.1   *A typical triangular finite element*

equation A.1 is that on the boundaries either the value of ψ must be prescribed, or its first derivative with respect to distance normal to the boundary must be zero. According to equations 5.60, these requirements are satisfied.

   Figure A.1 shows a typical triangular element, numbered m, in the solution domain. It has nodes at its corners numbered i, j and k, and dimensions as shown. Local coordinates x' and z' are parallel to x and z but have their origin at node i. Assuming a linear distribution of the stream function over the element

$$\psi(x', z') = C_1 + C_2 x' + C_3 z' \tag{A.2}$$

where C_1, C_2 and C_3 can be found in terms of the nodal-point values of the stream function, ψ_i, ψ_j and ψ_k, as follows

$$C_1 = \psi_i, \quad \begin{pmatrix} C_2 \\ C_3 \end{pmatrix} = \frac{1}{2\Delta_m}(B)\begin{pmatrix} \psi_i \\ \psi_j \\ \psi_k \end{pmatrix} = \frac{1}{2\Delta_m}(B)(\delta)_m \tag{A.3}$$

where Δ_m is the area of the element

$$\Delta_m = \frac{(a_k b_j - a_j b_k)}{2} \tag{A.4}$$

and (B) is a matrix of element dimensions

$$(B) = \begin{pmatrix} b_i & b_j & b_k \\ a_i & a_j & a_k \end{pmatrix} \tag{A.5}$$

Now, because the interelement boundaries make no contribution to the integral expressed in equation A.1

$$\frac{\partial \chi}{\partial \eta} = \sum_m \frac{\partial \chi^{(m)}}{\partial \eta} = 0 \tag{A.6}$$

where $\chi^{(m)}$ is the contribution of typical element m to the total value of χ. With the linear distribution of ψ over the element given by equation A.2, the following approximation may be used

$$\frac{\partial \chi^{(m)}}{\partial \eta} = \Delta_m \frac{\bar{\mu}}{\bar{H}^3} \left(C_2 \frac{\partial C_2}{\partial \eta} + C_3 \frac{\partial C_3}{\partial \eta} \right) \tag{A.7}$$

where $\bar{H}$ is the mean channel depth over the element.

The derivatives of $\chi^{(m)}$ with respect to the three nodal point values of ψ associated with element m may therefore be expressed as

$$\begin{vmatrix} \dfrac{\partial \chi^{(m)}}{\partial \psi_i} \\[2mm] \dfrac{\partial \chi^{(m)}}{\partial \psi_j} \\[2mm] \dfrac{\partial \chi^{(m)}}{\partial \psi_k} \end{vmatrix} = \frac{\Delta_m \bar{\mu}}{\bar{H}^3} \begin{vmatrix} \dfrac{\partial C_2}{\partial \psi_i} & \dfrac{\partial C_3}{\partial \psi_i} \\[2mm] \dfrac{\partial C_2}{\partial \psi_j} & \dfrac{\partial C_3}{\partial \psi_j} \\[2mm] \dfrac{\partial C_2}{\partial \psi_k} & \dfrac{\partial C_3}{\partial \psi_k} \end{vmatrix} \begin{pmatrix} C_2 \\ C_3 \end{pmatrix} = \frac{\bar{\mu}}{4\Delta_m \bar{H}^3} (B)^{\mathrm{T}}(B)(\delta)_m \tag{A.8}$$

where the superscript T indicates a matrix transposition. Combining equations A.6 and A.8

$$\sum_m \frac{\bar{\mu}}{4\Delta_m \bar{H}^3} (B)^{\mathrm{T}}(B)(\delta)_m = (K)(\delta) = 0 \tag{A.9}$$

where (δ) is a vector containing the stream function values for all the nodal points in the mesh. Square matrix (K) is the over-all viscous stiffness matrix, and contains coefficients assembled from the properties and dimensions of the individual elements.

Before equations A.9 can be solved for the unknown values of ψ, the boundary conditions defined by equations 5.60 must be imposed by appropriately modifying equations associated with boundary nodes at which the value of ψ is prescribed. The equations are not linear because the element mean viscosities, $\bar{\mu}$, are dependent on the local gradients of ψ. Using equations 5.48, 5.56 and A.2

$$\bar{\mu} = \mu_0 \left[\frac{(C_2{}^2 + C_3{}^2)^{1/2}}{\gamma_0 \bar{H}^2} \right]^{n-1} \tag{A.10}$$

An iterative method of solution of the Gauss–Seidel type may be used. The equations are first linearised by assuming suitable constant values for the element viscosities, and a few iterations are then performed to estimate the nodal point values of the stream function. Using these values to update the viscosities, the process is repeated until satisfactory convergence is achieved.

Appendix B

Solution of the Screw Channel Developing Melt Flow Equations

The purpose of this appendix is to explain in some detail suitable methods for solving equations 6.64 to 6.66, subject to the conditions defined by equations 6.69 and 6.70, together with the appropriate dimensionless temperature initial and boundary conditions. Together, these define the developing flow model for polymer melt flow in an extruder screw channel. Considering first the velocity analysis at a particular downstream position where the temperature profile is known, equations 6.64 and 6.65 may be integrated to give

$$\pi_P(Y^* - Y_0^*) = \frac{dW^*}{dY^*}(4I_2^*)^{(n-1)/2} \exp(-T^*) \tag{B.1}$$

$$\pi_X(Y^* - Y_1^*) = \frac{dU^*}{dY^*}(4I_2^*)^{(n-1)/2} \exp(-T^*) \tag{B.2}$$

where $\tau_{zy} = 0$ at $Y^* = Y_0^*$, and $\tau_{xy} = 0$ at $Y^* = Y_1^*$. These results may be rearranged to give

$$\frac{dW^*}{dY^*} = \pi_P(Y^* - Y_0^*)F(Y^*) \tag{B.3}$$

$$\frac{dU^*}{dY^*} = \pi_X(Y^* - Y_1^*)F(Y^*) \tag{B.4}$$

where, using equation 6.67, the function $F(Y^*)$ is given by

$$F(Y^*) = [(\pi_P Y^* - \pi_P Y_0^*)^2 + (\pi_X Y^* - \pi_X Y_1^*)^2]^{(1-n)/2n} \exp(T^*/n) \tag{B.5}$$

Hence, using the boundary conditions given by equations 6.69

$$W^* = \pi_P \int_0^{Y^*} (\alpha - Y_0^*)F(\alpha)\, d\alpha \tag{B.6}$$

$$1 = \pi_P \int_0^1 (\alpha - Y_0{}^*)F(\alpha) \, d\alpha = \pi_P(J_1 - Y_0{}^*J_0) \tag{B.7}$$

where

$$J_m = \int_0^1 \alpha^m F(\alpha) \, d\alpha \tag{B.8}$$

Equation B.6 may be integrated to find the dimensionless flow rate defined by equations 6.70

$$\pi_Q = \pi_P \int_0^1 \int_0^{Y^*} (\alpha - Y_0{}^*)F(\alpha) \, d\alpha \, dY^*$$

$$= \pi_P \int_0^1 (1 - \alpha)(\alpha - Y_0{}^*)F(\alpha) \, d\alpha$$

$$= \pi_P[(1 + Y_0{}^*)J_1 - J_2 - Y_0{}^*J_0] \tag{B.9}$$

Similar treatment of equation B.4 for transverse flow yields the following results, which are equivalent to equations B.7 and B.9

$$\tan \theta = \pi_X(J_1 - Y_1{}^*J_0) \tag{B.10}$$

$$0 = \pi_X[(1 + Y_1{}^*)J_1 - J_2 - Y_1{}^*J_0] \tag{B.11}$$

Equations B.7, B.9, B.10 and B.11, together with definitions B.5 and B.8, must be solved for the four unknowns π_P, $Y_0{}^*$, π_X and $Y_1{}^*$. As $Y_0{}^*$, which defines the position of the downstream flow stress neutral surface, can take any value between $-\infty$ and $+\infty$, it is more convenient to choose as an alternative variable the product $\pi_P Y_0{}^*$, which is much more limited in its range of possible values. Similarly, the product $\pi_X Y_1{}^*$ can be used in place of $Y_1{}^*$. Hence, the four equations may be rearranged for the new unknowns as

$$\pi_P = \beta(J_0 - J_1 - J_0\pi_Q) \tag{B.12}$$

$$\pi_P Y_0{}^* = \beta(J_1 - J_2 - J_1\pi_Q) \tag{B.13}$$

$$\pi_X = \beta(J_0 - J_1)\tan \theta \tag{B.14}$$

$$\pi_X Y_1{}^* = \beta(J_1 - J_2)\tan \theta \tag{B.15}$$

where

$$\beta = (J_0 J_2 - J_1{}^2)^{-1} \tag{B.16}$$

One method that has been used successfully to solve these simultaneous nonlinear algebraic equations is the Newton–Raphson technique. All four are of the general form

$$f_n(x_1, x_2, x_3, x_4) = 0, \quad n = 1, 2, 3, 4 \tag{B.17}$$

where $x_1 \equiv \pi_P$, $x_2 \equiv \pi_P Y_0{}^*$, $x_3 \equiv \pi_X$, $x_4 \equiv \pi_X Y_1{}^*$. For example

$$f_1 = \pi_P - \beta(J_0 - J_1 - J_0\pi_Q) = 0 \tag{B.18}$$

Given some initial estimates, $x_m{}^0$, for the unknowns, the corresponding function values, $f_n{}^0$, can be determined, and in general are not zero. Improved values of the x_m may be obtained with the aid of the Taylor series expansions of the f_n

$$f_n \approx f_n{}^0 + \sum_{m=1}^{4} h_m \frac{\partial f_n}{\partial x_m} = 0 \tag{B.19}$$

where the partial derivatives are evaluated at $x_m = x_m{}^0$, and the h_m are the estimated corrections to the x_m

$$x_m = x_m{}^0 + h_m \tag{B.20}$$

Equations B.19 can be rearranged in the form of four linear equations for the h_m

$$\left(\frac{\partial f_n}{\partial x_m}\right)(h_m) = -(f_n{}^0) \tag{B.21}$$

The square matrix of derivatives on the left-hand side of this result contains the partial derivatives of each of the f_n with respect to each of the four unknowns, a total of sixteen derivatives. These can be found numerically; for example

$$\frac{\partial f_1}{\partial x_2} = \frac{f_1(x_1{}^0, x_2{}^0 + \delta, x_3{}^0, x_4{}^0) - f_1(x_1{}^0, x_2{}^0, x_3{}^0, x_4{}^0)}{\delta} \tag{B.22}$$

where δ is some suitably small increment.

Having found the h_m from equations B.21, the improved x_m are obtained from equations B.20, and the whole iterative process repeated until there are no significant changes in the values of the unknowns between successive iterations. In practice, provided reasonably good initial estimates are available, the process converges satisfactorily in three or four iterations. For all but the first velocity analysis, good initial values are provided by the converged solutions from the analysis at the previous step along the channel. For the first velocity analysis, however, acceptable initial values are provided by the isothermal newtonian flow solutions described in section 6.1.2.

Turning now to the temperature analysis, equation 6.66 is first expressed in finite difference form. Figure B.1 shows part of a finite difference grid covering the channel cross-section illustrated in figure 6.8. The grid points at which the melt temperature is to be computed are distinguished by the counters i and j, i being used as a counter in the Z^* direction, and j in the Y^* direction. At the typical grid point i, j, approximations to the derivatives involved in the energy equation are

$$\frac{\partial T^*}{\partial Z^*} = \frac{T_{i+1,j}{}^* - T_{i,j}{}^*}{\delta Z^*} \tag{B.23}$$

$$C_{i,j} = \frac{\partial^2 T^*}{\partial Y^{*2}} = \frac{T_{i,j+1}{}^* - 2T_{i,j}{}^* + T_{i,j-1}{}^*}{(\delta Y^*)^2} \tag{B.24}$$

where $C_{i,j}$ represents the conduction term at the particular grid point. If $D_{i,j}$ is the corresponding dissipation term in equation 6.66

$$D_{i,j} = GS^{1-n}(4I_2{}^*)^{(n+1)/2} \exp(-T_{i,j}{}^*) \tag{B.25}$$

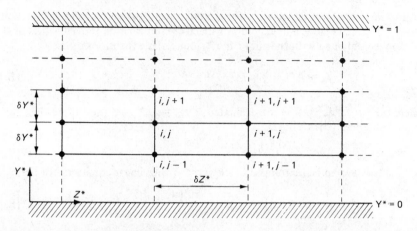

Figure B.1   *Part of a finite difference grid for computing temperature profiles*

then the equation may be expressed as

$$S_i{}^2 Pe W_{i,j}{}^* \frac{T_{i+1,j}{}^* - T_{i,j}{}^*}{\delta Z^*} = C_{i,j} + D_{i,j} \qquad (\text{B.26})$$

Now, because the $W_{i,j}{}^*$, $T_{i,j}{}^*$, $C_{i,j}$ and $D_{i,j}$ are all known, this equation can be used to find the $T_{i+1,j}{}^*$ at the new downstream position. In practice, however, this explicit approach to solving the parabolic partial differential energy equation is only stable for very small downstream increments, δZ^*. An implicit method of the Crank–Nicholson type, in which the conduction term is averaged between the old and new downstream positions, is much more useful. Thus, in place of $C_{i,j}$ in equation B.26, the following term is used

$$\frac{1}{2}\left[C_{i,j} + \frac{T_{i+1,j+1}{}^* - 2T_{i+1,j}{}^* + T_{i+1,j-1}{}^*}{(\delta Y^*)^2} \right] \qquad (\text{B.27})$$

The disadvantage of this implicit approach is that the unknown temperatures $T_{i+1,j+1}{}^*$ and $T_{i+1,j-1}{}^*$ are introduced into the equation for $T_{i+1,j}{}^*$. However, the equations for temperatures at grid points within the flow at the new downstream position, together with dimensionless forms of equations 6.8 and 6.58 for the temperatures at the barrel and screw surfaces, form a tridiagonal linear set and a well established direct method of solution exists for such a set.

Even using the implicit method of solving the energy equation, there is a limit to the size of the downstream increment for the process to be stable. This is of the order of

$$\delta Z^* < 4(\delta Y^*)^2 Pe \qquad (\text{B.28})$$

For typical values of $\delta Y^* = 0.05$ (that is, with 21 grid points over the depth of the channel, which is ample) and $Pe = 10^3$, δZ^* (that is, $\delta z/H_0$) should not exceed 10. This is rarely a significant limitation in practice.

Having computed the new temperature profile, the corresponding velocity profile, which will show only small changes from the previous one, can be obtained. Note that in evaluating the J_m defined by equation B.8, numerical integration of products involving the function $F(Y^*)$ given by equation B.5 must be performed. This is because the required temperature profile is only defined at the finite difference grid points. Simpson's rule provides a simple and accurate method of integration.

Appendix C

Solution of the Melting Model Equations

The purpose of this appendix is to describe a method for solving differential equations 6.99, 6.102 and 6.106, together with mass and force balance equations 6.103 and 6.104, governing downstream development of the melting model variables $\delta_1, \delta_2, V_{sz}, X$ and P_z. The three differential equations are of the form

$$\frac{dy_i}{dz} = f_i(x_1, x_2, x_3), \quad i = 1, 2, 3 \tag{C.1}$$

where $x_1 \equiv \delta_1, x_2 \equiv \delta_2, x_3 \equiv V_{sz}$ and

$$f_1 = \omega_1 X - (m_{1x} - m_{fx}) \tag{C.2}$$

$$f_2 = \omega_2(X + H_s) \tag{C.3}$$

$$f_3 = -\omega_1 X - \omega_2(X + H_s) \tag{C.4}$$

$$y_i = g_i(x_1, x_2, x_3) \tag{C.5}$$

Expressing the y_i functions explicitly

$$y_1 = m_{1z}X, y_2 = m_{2z}(X + H_s), y_3 = \rho_s V_{sz} X H_s \tag{C.6}$$

Now, provided the downstream distance, δz, between successive channel cross-sections is small, the values of these mass flow rates at the new cross section, denoted by $y_i{'}$, can be obtained from those at the previous section by expressing equations C.1 in approximate finite difference form

$$y_i{'} = y_i + f_i(x_1, x_2, x_3)\delta z, \quad i = 1, 2, 3 \tag{C.7}$$

Since these new values must satisfy equations C.5 with the new values of δ_1, δ_2 and V_{sz} denoted by $x_i{'}$

$$y_i{'} = g_i(x_1{'}, x_2{'}, x_3{'}), \quad i = 1, 2, 3 \tag{C.8}$$

which represents a set of nonlinear algebraic equations. These equations can be solved for δ_1, δ_2 and V_{sz} by a Newton–Raphson method of the type described in appendix B. In order to evaluate the g_i during this iterative procedure, it is necessary to find both X and P_z, using equations 6.103 and 6.104, respectively.

Further Reading

The following list contains a selection of published books and papers concerned with various aspects of the principles of polymer processing. The first general section contains books that offer either detailed descriptions of equipment and processes or analytical treatments of them. Subsequent sections list publications that are relevant to particular topics covered in the chapters of this book. In the general section, books are listed alphabetically by author's name, while under chapter headings items are arranged in the order in which the topics occur in the text.

General

Bernhardt, E. C., *Processing of Thermoplastic Materials* (Reinhold, New York, 1974).

McKelvey, J. M., *Polymer Processing* (Wiley, New York, 1962).

Middleman, S., *Fundamentals of Polymer Processing* (McGraw-Hill, New York, 1977).

Ogorkiewicz, R. M., *Thermoplastics: Effects of Processing* (Iliffe, London, 1969).

Pearson, J. R. A., *Mechanical Principles of Polymer Melt Processing* (Pergamon, Oxford, 1966).

Chapter 2

Fisher, E. G., *Extrusion of Plastics*, 3rd ed. (Newnes-Butterworth, London, 1976).

Schenkel, G. P. M., *Plastics Extrusion Technology and Theory* (Iliffe, London, 1966).

Janssen, L. P. B. M., *Twin Screw Extrusion* (Elsevier, Amsterdam, 1978).

Rubin, I. I., *Injection Molding: Theory and Practice* (Wiley-Interscience, New York, 1973).

Pye, R. G. W., *Injection Mould Design*, 2nd ed. (George Godwin, London, 1978).

Fisher, E. G., *Blow Moulding of Plastics* (Iliffe, London, 1971).

Elden, R. A., and Swan, A. D., *Calendering of Plastics* (Iliffe, London, 1971).

Ziabicki, A., *Fundamentals of Fibre Formation* (Wiley, New York, 1976).

Chapter 3

Brydson, J. A., *Flow Properties of Polymer Melts* (Iliffe, London, 1970).
Wilkinson, W. L., *Non-Newtonian Fluids* (Pergamon, Oxford, 1960).
Harris, J., *Rheology and Non-Newtonian Flow* (Longman, London, 1977).
Walters, K., *Rheometry* (Chapman and Hall, London, 1975).
Van Wazer, J. R., Lyons, J. W., Kim, K. Y., and Colwell, R. E., *Viscosity and Flow Measurement* (Interscience, New York, 1963).
Lodge, A. S., *Elastic Liquids* (Academic, London, 1964).

Chapter 4

Astarita, G., and Marrucci, G., *Principles of Non-Newtonian Fluid Mechanics* (McGraw-Hill, London, 1974).
Bird, R. B., Stewart, W. E., and Lightfoot, E. N., *Transport Phenomena* (Wiley, New York, 1960).
Serrin, J., 'The Derivation of Stress–Deformation Relations for a Stokesian Fluid', *J. Math. Mech.* 8 (1959) p. 459.
Richardson, S., 'On the No-slip Boundary Condition', *J. Fluid Mech.*, 59 (1973). p. 707.
Pearson, J. R. A., 'Heat Transfer Effects in Flowing Polymers', in *Progress in Heat and Mass Transfer*, vol. 5, ed. W. R. Schowalter *et al.* (Pergamon, New York, 1972) p. 73.
Pearson, J. R. A., 'Variable-viscosity Flows in Channels with High Heat Generation', *J. Fluid Mech.* 81 (1977) p. 191.
Pearson, J. R. A., 'Polymer Flows Dominated by High Heat Generation and Low Heat Transfer', *Polym. Eng. Sci.*, 18 (1978) p. 222.
Pearson, J. R. A., 'The Lubrication Approximation Applied to Non-Newtonian Flow Problems: A Perturbation Approach', in *Proceedings of the Symposium on Solution of Non-linear Partial Differential Equations*, ed. W. F. Ames (Academic, New York, 1967) p. 73.
Benis, A. M., 'Laminar Flow of Power-law Fluids through Narrow Three-dimensional Channels of Varying Gap', *Chem. Engng. Sci.*, 22 (1967) p. 805.
Funt, J. M., *Mixing of Rubber* (Rubber and Plastics Research Association, 1976).
Bigg, D. M., 'On Mixing in Polymer Flow Systems', *Polym. Eng. Sci.*, 15 (1975) p. 684.

Chapter 5

Pearson, J. R. A., and Petrie, C. J. S., 'A Fluid-mechanical Analysis of the Film-blowing Process', *Plast. Polym.*, 38 (1970) p. 85.
Pearson, J. R. A., and Petrie, C. J. S., 'The Flow of a Tubular Film. Part I. Formal Mathematical Representation', *J. Fluid Mech.*, 40 (1970) p. 1.
Pearson, J. R. A., and Petrie, C. J. S., 'The Flow of a Tubular Film. Part 2. Interpretation of the Model and Discussion of Solutions', *J. Fluid Mech.*, 42 (1970) p. 609.
Matovich, M. A., and Pearson, J. R. A., 'Spinning a Molten Threadline: Steady State Isothermal Viscous Flows', *Ind. Eng. Chem. Fund.*, 8 (1969) p. 512.
Pearson, J. R. A., 'Non-Newtonian Flow and Die Design: I General Principles and

Specification of Flow Properties; II Flow in Narrow Channels of Pseudoplastic Fluids', *Trans. Plast. Inst., Lond.,* 30 (1962) p. 230.

Pearson, J. R. A., 'Non-Newtonian Flow and Die Design: III A Cross-Head Die Design', *Trans. Plast. Inst., Lond.,* 31 (1963) p. 125.

Pearson, J. R. A., 'Non-Newtonian Flow and Die Design: IV Flat-film Die Design', *Trans. Plast. Inst., Lond.,* 32 (1964) p. 239.

Fenner, R. T., and Williams, J. G., 'Analytical Methods of Wire-coating Die Design', *Trans. Plast. Inst., Lond.,* 35 (1967) p. 701.

Caswell, B., and Tanner, R. I., 'Wirecoating Die Design Using Finite Element Methods', *Polym. Eng. Sci.* 18 (1978) p. 416.

Fenner, R. T., and Nadiri, F., 'Finite Element Analysis of Polymer Melt Flow in Cable-covering Crossheads', *Polym. Eng. Sci.,* 19 (1979) p. 203.

Fenner, R. T., *Finite Element Methods for Engineers* (Macmillan, London, 1975).

Kiparissides, C., and Vlachopoulos, J., 'Finite Element Analysis of Calendering', *Polym. Eng. Sci.,* 16 (1976) p. 712.

Martin, B., 'Some Analytical Solutions for Viscometric Flows of Power-law Fluids with Heat Generation and Temperature Dependent Viscosity', *Int. J. Non-linear Mechanics,* 2 (1967) p. 285.

Pearson, J. R. A., 'On the Melting of Solids near a Hot Moving Interface, With Particular Reference to Beds of Granular Polymers', *Int. J. Heat Mass Transfer,* 19 (1976) p. 405.

Chapter 6

Fenner, R. T., *Extruder Screw Design* (Iliffe, London, 1970).

Fenner, R. T., 'Developments in the Analysis of Steady Screw Extrusion of Polymers', *Polymer,* 18 (1977) p. 617.

Griffith, R. M., 'Fully Developed Flow in Screw Extruders', *Ind. Engng. Chem. Fund.,* 1 (1962) p. 180.

Zamodits, H. J., and Pearson, J. R. A., 'Flow of Polymer Melts in Extruders. Part I. The Effect of Transverse Flow and of a Superposed Steady Temperature Profile', *Trans. Soc. Rheol.,* 13 (1969) p. 357.

Carlile, D. R., and Fenner, R. T., 'On the Lubricating Action of Molten Polymers in Single-screw Extruders', *J. Mech. Eng. Sci.,* 20 (1978) p. 73.

Darnell, W. H., and Mol, E. A. J., 'Solids Conveying in Extruders', *S.P.E. J.,* 12 (1956) p. 20.

Lovegrove, J. G. A., and Williams, J. G., 'Solids Conveying in a Single Screw Extruder: The Rôle of Gravity Forces', *J. Mech. Eng. Sci.,* 15 (1973) p. 114.

Lovegrove, J. G. A., and Williams, J. G., 'Solids Conveying in a Single Screw Extruder: A Comparison of Theory and Experiment', *J. Mech. Eng. Sci.,* 15 (1973) p. 195.

Lovegrove, J. G. A., and Williams, J. G., 'Pressure Generation Mechanisms in the Feed Section of Screw Extruders', *Polym. Eng. Sci.,* 14 (1974) p. 589.

Lovegrove, J. G. A., 'An Examination of the Stresses in Solids Being Conveyed in a Single-screw Extruder', *J. Mech. Eng. Sci.,* 16 (1974) p. 281.

Tadmor, Z., and Klein, I., *Engineering Principles of Plasticating Extrusion,* (Van Nostrand-Reinhold, New York, 1970).

Edmondson, I. R., and Fenner, R. T., 'Melting of Thermoplastics in Single Screw Extruders', *Polymer,* 16 (1975) p. 49.

Shapiro, J., Halmos, A. L., and Pearson, J. R. A., 'Melting in Single Screw Extruders; Part I: The Mathematical Model; Part II: Solution for a Newtonian Fluid', *Polymer,* 17 (1976) p. 905.

Fenner, R. T., Cox, A. P. D., and Isherwood, D. P., 'Surging in Screw Extruders', *S.P.E. Annual Technical Conference* (1978) p. 494.

Pearson, J. R. A., 'On the Scale-up of Single-screw Extruders for Polymer Processing', *Plastics and Rubber: Processing,* 1 (1976) p. 113.

Yi, B., and Fenner, R. T., 'Scaling-up Plasticating Screw Extruders on the Basis of Similar Melting Performance', *Plastics and Rubber: Processing,* 1 (1976) p. 119.

Fenner, R. T., 'The Design of Large Hot Melt Extruders', *Polymer,* 16 (1975) p. 298.

Chapter 7

Nunn, R. E., and Fenner R. T., 'Reciprocating-screw Plastication', *S.P.E. Annual Technical Conference* (1978) p. 72.

Donovan, R. C., 'The Plasticating Process in Injection Molding', *Polym. Eng. Sci.,* 14 (1974) p. 101.

Nunn, R. E., and Fenner, R. T., 'Flow and Heat Transfer in the Nozzle of an Injection Moulding Machine', *Polym. Eng. Sci.,* 17 (1977) p. 811.

Williams, G., and Lord, H. A. 'Mold-filling Studies for the Injection Molding of Thermoplastic Materials; Part 1: The Flow of Plastic Materials in Hot- and Cold-walled Circular Channels; Part 2: The Transient Flow of Plastic Materials in the Cavities of Injection-moulding Dies', *Polym. Eng. Sci.,* 15 (1975) p. 553.

Stevenson, J. F., 'A Simplified Method for Analyzing Mold-filling Dynamics. Part 1: Theory', *Polym. Eng. Sci.,* 18 (1978) p. 577.

Index

additives 2, 15
adiabatic flow 27, 148, 151
annular flow 7–8, 53, 62–6
apparent shear rate and viscosity 25, 27
axisymmetric flow 6, 54, 153

barrel, of capillary 22, 23, 26, 29
    of extruder 4, 5, 6, 10, 93, 94, 95,
        114, 115, 116, 123, 124, 125
barrel temperature 5, 53, 97, 111, 113,
    142
barrel velocity 95, 99, 100, 101, 126
blow moulding 4, 13, 53
body forces 36, 44, 94, 95, 122
boundary condition 30, 36, 43, 47, 68,
    73, 109, 123
        no-slip 27, 28, 43, 55, 59, 63, 64,
           96
        pressure 73, 81, 82, 150
        thermal 43, 46, 47, 88, 91, 97,
           109, 110, 126, 140, 151
        velocity 25, 43, 46, 64, 80, 81, 89,
           96, 102, 103, 104, 106, 110, 111
breaker plate 5, 139
break-up of solid bed 125, 134–6, 138
Brinkman number 46, 47, 86, 90, 140

cable covering 6, 9, 66, 71–4, 77–9
calendering 4, 13–15, 53, 79–85
capillary rheometer 20, 22–32, 40, 54, 66
chill roll casting 7
coat-hanger die 6–7, 75–7
cold slug well 12
compounding 2
compressibility 31
compression ratio 6, 120, 141
compression section 5, 110, 125, 134, 141
cone-and-plate rheometer 20–2, 25, 28–
    30

constitutive equation 36, 37–42, 45, 52,
    54, 63, 70, 96
continuity equation 36, 56, 67, 95
continuum mechanics 33, 35–7, 53, 93,
    115
control of processes 3, 5, 53, 75, 113–14,
    115, 138, 139
critical stress 31, 43, 115
cross viscosity 38, 40, 41, 42, 45, 96
crosshead dies 8–9, 13, 66–8, 71–4,
    77–9, 158

deflector in crosshead die 9, 71–4, 77–9
degradation 2, 6, 15, 16, 22, 29, 142, 144,
    147
delay zone 123
density 23–4, 27, 31, 36, 39, 40, 42, 95,
    116, 135
design 3, 44, 53, 80
    of dies 6, 71, 74–9
    of moulds 153, 158
    of screws 109, 117, 138, 139–44
developing flow 48, 49, 54, 97, 99, 108,
    109–14, 130, 141–4, 151, 163–7
die 2, 4, 6–9, 22, 33, 44, 53–79, 139
die face cutting 2
die swell 29–31
dilatant materials 18
dimensional analysis 43–8, 99–102, 140,
    141, 147–9

ejector pin 12
elasticity of polymers 2, 6, 8, 13, 16, 29–
    31, 38, 41, 48–9, 96
elastomers 1
energy conservation equation 36, 37, 43,
    45, 47–8, 49, 96, 99, 109, 149, 158
enthalpy 16–17, 132
equilibrium, equation of 36–7, 43, 47

extensional flow   41–2
extrudate   4, 6, 7, 24, 29, 30, 68
extrusion   4
extrusion coating   7

feed angle   116–21
feed pocket   5, 133
feed section   5, 6, 11, 115–23, 142
feeding in extruders   4–5, 32, 115–23, 139, 142
feedstock   5, 115, 123, 124
filaments   3, 15, 42
film, extruded   3, 6–7, 8–9, 33
    flat   6–7, 66, 75–7
    gate   155
    thin sheared   53, 85–92, 114, 121, 125, 123–33, 145–6
    tubular   8–9, 15, 41, 53, 66
finite-difference method   68, 109, 112, 130, 132, 133, 165–7
finite-element method   68, 71, 73, 109, 160–2
flight lead   94, 119
flight of an extruder screw   5, 94–5, 96, 98, 109, 114–15, 116, 117, 118, 122, 123, 125
flight pitch   94, 134
flight width   94
flow, curve   18–19, 28–9, 31, 107–8
    dimensionless rate   64–5, 89, 100, 103–4, 107, 115, 117, 142, 164
    drag   81, 85–92, 102–3, 104–5, 107, 114–15, 128, 131, 137, 139, 142
    free surface   3, 8, 30, 41, 43, 53
    fully developed   48–9, 65, 73, 97, 98, 99, 108–9, 110, 126
    narrow channel   66–74, 158, 160–2
    one-dimensional   66, 75, 99, 105, 117, 121, 122, 123
    pressure   102–3, 104–5, 107, 137, 151
    rate   23, 25, 30–1, 56, 59, 67, 75–6, 81, 99, 103–5, 106, 114–15, 116, 128, 131, 139
    recirculating   80, 103, 109, 125, 137
    thin film   85–92
    three-dimensional   66, 70, 108
    two-dimensional   66, 75, 80, 99, 108, 122–3, 158
friction   32, 116–23

gate to a mould cavity   11–13, 152, 155–7
grade of material   1, 16, 65
Graetz number   46–8, 49, 73, 84, 100, 140–1
granules   2, 4, 20, 32, 85, 115, 116, 119, 124

Griffith number   46–8, 54, 73, 84, 100, 111, 140–1

haul off   8, 53
helix angle   94, 100–1, 117, 119, 120, 140
homogeneity   29, 38, 49, 52
homogenisation   2, 4, 6, 110, 141–4
hopper   5, 10, 32, 115, 121, 139

inertia effects   30, 36, 44, 47, 48–9, 65, 94, 96
initial condition   46, 130, 156, 159, 163
injection moulding   3, 4, 9–13, 31, 44, 145–59
invariants of rate-of-deformation tensor   38–40, 41, 45, 51, 96, 137
isothermal flow   47, 54, 55, 56, 59, 60, 62, 64, 66, 90, 91, 99, 105, 108, 126, 130, 136, 150, 153

latent heat of fusion   17, 91, 126, 132, 159
leakage flow   96–7, 114–15, 129, 136, 137
length-to-diameter ratio   6, 11, 26, 140, 142, 143
lubrication approximation   48–9, 56, 60, 66, 70, 80, 81, 97, 98, 109

manifold of sheet die   7, 75–7
mass conservation equation   36, 110–11
melt   2
melt fed extruder   93, 110, 139, 141–4
melt flow index   22–3, 26, 29
melt fracture   30–1, 65
melt pool   124–5, 130–1, 146
melt properties   16–32
melting, in extruders   4–5, 85, 89, 90–2, 93, 123–36, 138, 141, 168
    in injection moulding machines   9–10, 145–7
melting point temperature   16–17, 85, 91, 124, 126, 132, 133, 158
metering section   5, 6, 103, 105, 109, 115, 139
mixing   2, 11, 15, 49–52
    in calendering   14, 80, 84–5
    in extruders   5, 6, 128, 137–8, 140, 141–4
molecular weight   1, 2, 15
molecular weight distribution   2, 15
momentum conservation equation   36–7, 44, 95, 149
mould   9–13, 152–9
multiscrew extruders   6, 10

Navier–Stokes equations   36
newtonian flow   18, 19, 21, 25, 29–30, 40, 41, 42, 55–6, 64, 81–3, 99, 102–5, 130, 136

nip between calender rolls 14, 79, 80, 82, 84
nonisothermal flow 85, 87–90, 92, 153
non-newtonian flow 2, 17–19, 21, 25, 27, 38–41, 55–6, 83–4, 87–90, 105–14
normal stress 41, 43
nozzle 10, 11, 19, 145, 147–51
nylon 1, 19

orientation of molecules 15, 22, 152

parison 13
Peclet number 46–7, 84, 100, 109, 140
pipe extrusion 7–8, 66, 68–71
plastication 4, 115, 145–6
plastics 1, 22
plug flow 28, 32, 116–23, 151
polyethylene 1, 19, 22, 32, 57, 61, 65, 141, 147, 149, 151
polymer manufacture 2, 6
polymeric materials 1–2
polymerisation 2, 4
polypropylene 1, 19, 32
polystyrene 1, 32, 100, 112, 133, 134
polyvinyl chloride 1, 13, 19, 28, 125
powder 2, 4, 10, 32, 85, 115, 125
power consumption 82, 84, 115, 136–7, 140, 141–2
power-law, constitutive equation 18–19, 27, 28, 31, 41, 42, 55, 63, 70, 87–90, 105
pressure, definition of 35
        effect on properties 16, 17, 19, 25, 26, 31–2, 41, 145, 147, 149–51
pressure coefficient of viscosity 32, 149
processing, effects of 15
        range of shear rates 18–19, 28–9
processing properties 16–32
processing temperature 15, 16, 71
pseudoplastic materials 18

quality 9, 11, 13, 49, 128, 140, 141

Rabinowitsch correction 27
rate-of-deformation tensor 34, 35, 38, 51, 96
rate-of-rotation tensor 35, 38
recirculating flow 80, 103, 109, 125, 137
residence time 30, 51, 148–9
Reynolds number 46–7, 49, 65, 84, 96, 100
rolls, calender 13–14, 79–85
rubbers 1, 4
runner 12, 152

sandwich moulding 13
scalar quantities 33–4, 38
scaling up 43–4, 140–1

screen pack 5, 139
screw extrusion 4–6, 49, 53, 93–144
screw injection moulding 10–11, 145–7
screwback 145–6
segregation, intensity and scale of 49–50
shape factors 104–5, 130
shear flow 17–18, 21, 22, 37, 39, 40–1, 42, 45, 50–1
shear rate 18, 21, 25, 27, 41, 46, 67, 84, 87, 99, 100, 131
shear strain 29, 51, 138
shear stress 15, 18, 21, 24–5, 27, 40, 46, 87, 100
sheet extrusion 6–7
slip 27, 28, 30, 43, 65, 96, 114, 116, 137
solid bed 85, 124–36, 145–6
        break-up of 125, 134–6, 138
solid conveying 115–23, 136, 138, 139, 141
solid polymer properties 32
specific heat 17, 27, 32, 37, 96, 126
spider in pipe die 8, 68–71
spinning of fibres 15, 41, 53
sprue 12, 152
stability of processes 3, 123, 135, 136, 138
steady flow 3, 21, 37, 44, 49, 56, 96, 129, 138, 145, 149, 151
stokesian fluid 38–40, 41, 45, 54, 96
strain 29, 38, 51, 138
stream function 67, 70, 73, 160–2
streamline 15, 29, 68–70, 78, 109
stress, critical 31, 43, 115
        neutral surface 63, 83, 105, 107, 164
        normal 41, 43
        shear 18, 34, 35, 46, 56, 60, 87, 100, 111, 115, 136, 155
        tensile 34, 41–2
        tensor 34
        total 34–5, 42
        viscous 34–5
surging in extruders 96, 123, 138, 146
swelling ratio 29–31

Tadmor model of melting 125–8, 134
temperature, bulk mean 112–13, 131, 132, 142
        dimensionless 45, 87, 111
temperature coefficient of viscosity 32, 45, 151
temperature effect on properties 16–17, 19, 31–2, 38, 47, 60, 147, 151
tensor, notation 33–5, 36
        rate-of-deformation 34, 35, 38, 51, 96
        rate-of-rotation 35, 38
        total stress 34–5
        velocity gradient 35, 38
        viscous stress 34–5

thermal conduction 2, 37, 47–8, 80, 85,
90, 93, 97, 109, 110, 124, 125, 130, 142,
145–6, 152, 158
thermal conductivity 13, 17, 32, 37, 96,
126, 130, 159
thermal contact 43, 61, 97
thermal convection 37, 46, 47–8, 66, 73,
97, 99, 109, 132, 142, 149
thermoforming 15
thermoplastics 1–3, 4, 6, 8, 15, 19
thermosetting materials 1, 4
time dependence of properties 1, 16, 32
torpedo (mandrel) in pipe die 7–8, 68–71
Trouton viscosity 42
twin-screw extruders 6, 10
two-stage screw 6

vector notation 33–4
velocity, dimensionless 45, 87, 111
gradient tensor 35, 38

mean 45, 56, 84, 148
relative 21, 44, 45, 85, 95, 116–
17, 126–7, 129, 131, 134–5
viscoelasticity 2, 16, 38, 138
viscosity, apparent 25, 27
cross 38, 40, 41, 42, 45, 96
definition of 18, 33, 38
effective 41
generalised 38, 40
mean 46, 67, 100, 131
measurement of 20–9, 145
pressure dependence 31–2, 96, 149,
150–1
shear rate dependence 17–19, 38–
40
temperature dependence 32, 45, 151

wear in extruders 3, 5, 144
wire covering 6, 9, 64–6